WORKING
ON THE
EDGE

WORKING
ON THE
EDGE

SURVIVING IN THE WORLD'S
MOST
DANGEROUS PROFESSION:
KING CRAB FISHING ON ALASKA'S
HIGH SEAS

SPIKE WALKER

ST. MARTIN'S PRESS NEW YORK

Design by Dawn Niles

Library of Congress Cataloging-in-Publication Data

Walker, Spike
    Working on the edge / Spike Walker.
        p.   cm.
    ISBN 0-312-06002-5
    1. Alaskan king crab fisheries—Alaska.   I. Title.
SH380.45.U5W35   1991
639'.544—dc20   90-27590
                                                                        CIP

First Edition: June 1991

10 9 8 7 6 5 4 3 2 1

To my father, Robert B. Walker. And to the families of the hundreds of fishermen who have died at sea in Alaska.

In the hope that the youthful tide adventuring north each year may know the perils awaiting them. And that the slaughter may end.

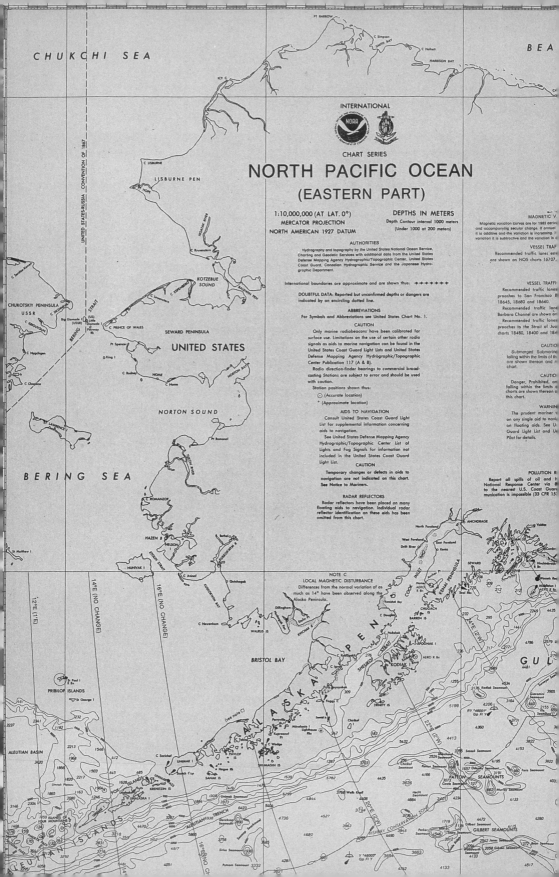

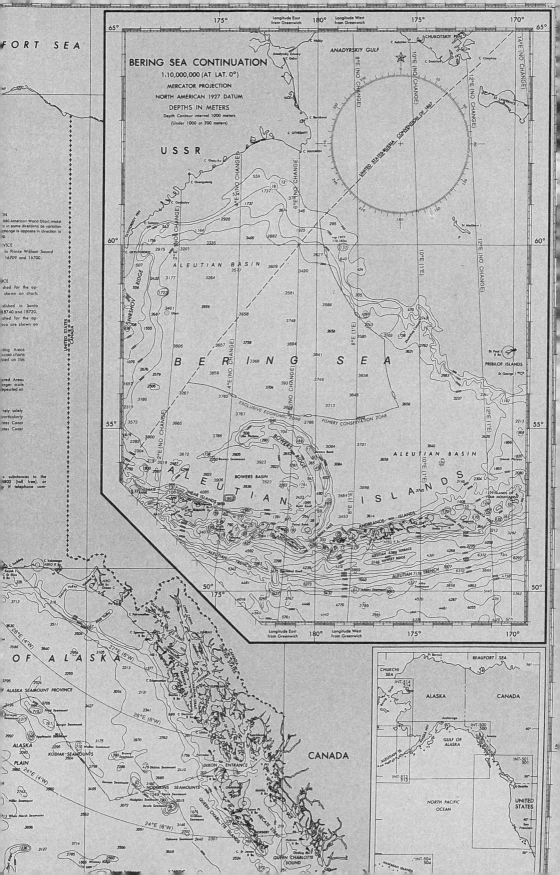

CONTENTS

ACKNOWLEDGMENTS

I would like to express my deep sense of gratitude to the hundreds of people who have contributed, in one form or another, to the creation of this book.

Foremost, there are my parents, Lorna M. Walker and, my late father, Robert B. Walker, who died during the final editing of the manuscript. Without their unwavering belief and support, its completion would not have been possible.

I would also like to thank Mike Wilson, the skipper of the 128-foot, Kodiak-based crab boat *Alaska Trojan,* for saving me a position with his halibut crew each season. Those profitable adventures provided the much-needed funds to keep this project going. They also provided welcome breathers during the years of research, and served as brutal reminders of the rugged physical existence and raw, wilderness world that I was attempting to recreate.

I must also thank the skippers I worked *for,* and the crewmen I worked *with* on board the crab fishing boats *Royal Quarry, Williwaw Wind, Rondys, Elusive,* and *Buccaneer,* as well as the *Alaska Trojan.* To my delight, almost without exception, these fishermen agreed to sit down and, with tape recorder rolling, share with me their memories of those days, as well as areas of their lives about which I knew nothing—and had no right to ask.

And this was true of untold numbers of cannery workers, cannery foremen, skippers, deckhands, U.S. Coast Guard SAR (Search & Rescue) helicopter pilots and investigators, as well as scores of shipwreck survivors, and, far too often, the relatives of fishermen who, crewing aboard one crab boat or another, rode out the mouth of St. Paul Harbor (Kodiak) never to return. And there were marine biologists, and Alaska State Troopers, Kodiak City Police Officers, as well as bartenders, church pastors, photographers, fishermen's wives, commercial divers, bush pilots, and several who, while sharing their experiences, wished to remain anonymous.

In alphabetical order, I would like to thank the following: Ken Alread, Bill and Jeanie Alwert; Debbie Schiedler Aranda; Fred Ball; Lt. Rick Bartlett (USCG); Chris Blackburn; Kelly Bledsoe; Bob Bronzon; Steve Calhoun; Dave Capri; Tom Casey; Bill Caulkins; Donny Channel; Sgt. Tom Culbertson; Mike Doyle; Oscar and Peggy Dyson; Bart and Toni Eaton; Larry Garrison; Robbie Green; Big Steve Griffing; Clifford Hall; John Hall; Pastor Hugh Hall and his wife, Margaret; Robbie Hall; Vern and Debbie Hall; Wilburn and Joyce Hall; Ole and Mary Harder; Bill Jacobson; Connie Johnson; George Johnson; Mike Jones; Kelly and Patty Joy; Blake Kinnear, Jr.; Pete Knudsen; Ray LeGrue; Lt. Bob Lockman (Alaska State Trooper); Loui Lowenberg; Nancy Luther; Tom Luther; Tom MacDonald; John Magoteaux; Richard Majdic; Les Martin; Corkey McCorkle; Mike McSwain; Jeanie Miller; Lt. Jimmy Ng (USCG); Mike and Sue Nolan; Danny O'Malley; Steve Oswald; Teddy Painter, Sr.; Teddy Painter, Jr.; Bert Parker, Jr.; Hank Pennington; Gerry Perry (USCG); Doug Petersen; Rudy Peterson; Kris Poulsen; Guy Powell; Mike Roust; Lance Russel; Terry and Barbara Sampson; Jim Sandlin; Buster Schaishnikoff; Larry Schaishnikoff; Wayne Schueffley; Charles "Rusty" Slayton; Commander Allen Steinman (USCG); Joe Szert; Jim Taylor; Tom Thiesson; Wallace Thomas; Richard Thummel; Nell Waage; Susey Wagner; Lt. David Watkins (USCG Survival Training Expert); Lt. Commander Tom Walters (USCG); Gary Wiggins; Rick Williams; Mike and Linda Wilson; and Dave Woodruff.

I will never forget the hospitality shown by Mrs. Nancy Freeman, owner and publisher of the *Kodiak Daily Mirror*, who allowed me open access to entire years of back issues of her newspaper. Mrs. Marion Johnson and the friendly and knowledgeable staff at Kodiak's fine Historical Society also proved to be most accommodating. In Anchorage, I was assisted by Lt. Jody Klimas and Lt. Brendt Whitner at the U.S. Coast Guard office. And there were long and wonderful days of pampered research in the Rare Books section of Ketchikan's Public Library. At the National Archives and Record Center in Seattle, Joyce Justice proved to be an invaluable resource. And in Seattle, John Pappenheimer, owner and publisher of *Alaska Fisherman's Journal*, was also extremely helpful.

In the world of letters, I would be remiss if I neglected to thank Brad Matsen, the Pacific editor at *National Fisherman* magazine. His fine editorial skills and unwavering encouragement in the early days when this book was only a dream were greatly appreciated.

I would be equally negligent if I did not thank such literary confidants as Bud Gardner, that dynamic speaker, motivator, and professor of English at American River College in Sacramento, California; as well as Mrs. Ethel "Mom" Bangert, retired professor of English, award-winning author, guest speaker, and writer also in Sacramento; and Mr. Duane Newcomb, former professor of English, author, guest speaker, and writer from Grass Valley, California; as well as the late Francis L. "Floyd" Jones, former professor of English and literary sage at Spokane Falls Community College in Spokane, Washington. And Jack Hart, former editor of *Northwest Magazine,* lecturer, and professor of journalism at the University of Oregon (Eugene).

Finally, I must thank Barbara Anderson, my talented senior editor at St. Martin's Press. Her original vision for this book proved prophetic, and her literary competence and intregrity, as well as her optimism, sustained me throughout this lengthy project.

INTRODUCTION

This book is about the most dangerous occupation on earth: king crab fishing off the coast of Alaska. It describes the eight-year period (1976 to 1984) that will forever mark the peak of the boom days when hundreds of naïve and ambitious young people like myself rushed north in the hope of landing a job and cashing in on the biggest bonanza in the history of Alaska. This eight-year span was no less than a modern-day gold rush, which netted fishermen more than $1.4 billion in profits.

During those turbulent and exciting years, three elements came together as they never had before. First, there was a sudden and inexplicable explosion in the numbers of harvestable king crab in the Bering Sea. Second, due to increased size and technological advances in gear, the catching capability of the new crab boats plying the waters of the Gulf of Alaska (out of the port of Kodiak) and in the Bering Sea (out of the port of Dutch Harbor in the Aleutian Islands) doubled, and then doubled again. At the same time, the domestic market for king crab exploded, and the price that canneries offered fishermen for their crab suddenly skyrocketed. Overnight, the waterfront ports of Kodiak and Dutch Harbor became boomtowns. And in the bars, anything could, and often did, happen.

When I arrived in Kodiak in the winter of 1978, the rush had already begun. And though I had no idea what lay ahead, I set out to secure a berth as a crewman aboard one of the crab boats in the local fleet. Over the next year, sixty-five crewmen were injured at sea seriously enough to require medical treatment. Some were evacuated by U.S. Coast Guard helicopters. Fourteen crewmen died outright at sea.

Why would anyone choose to venture into such a world, and take such risks? There are many reasons. This book will try to convey to you the sense of pride, freedom, and adventure that deckhands felt as they worked aboard crab boats plying Alaska's lonely and deadly wilderness of sea. It will also describe perhaps the most powerful lure of all: the very real possibility of landing a crewman's berth aboard one of the elite, highline crab boats and taking home a $100,000 paycheck in a single two-month season of king crabbing.

In addition to my eight years spent as a crewman aboard some of the largest and most successful crab boats in the American fleet, this project has involved five years of research and writing, including several months of work and wonderful discovery at the United States Coast Guard Files in Anchorage and Juneau, and at the National Archives in Seattle. The research has led me through dozens of books on the region, and (by telephone, car, and plane) has led me all over the United States, from Dutch Harbor, Alaska, to St. Augustine, Florida, as I tracked down and interviewed at least one survivor or participant from every significant event that occurred during that period from 1976 to 1984. I have talked at length with scores of key crewmen and skippers, and have gained invaluable background information from hundreds of other fishermen, wives, cannery workers, and support personnel who were involved during that period in the king crab industry.

PROLOGUE

Packing a moderate load of crab-pot gear secured to her back deck, and following a week of fishing near Icy Bay, the crab boat *Master Carl* was en route back to its home port of Cordova and was closing on the famous outcropping of rock known as Cape St. Elias when things began to come apart.

A parted weld in the hull, a malfunctioning pump system, a leaking hatch cover—no one would ever be able to tell exactly what went wrong, but somewhere off Prince William Sound's Montague Island, at about midnight on April 28, 1976, the ship began taking on water. At the first sign of a list, seasoned skipper Tom Miller of Cordova, Alaska, turned his ship to starboard and began running in the same direction as the swelling seas on a westward course into Alaska's Prince William Sound. He then ordered his crew to dump their load of crab pots stacked on the back deck. John Magoteaux, then of Othello, Washington, and Tom Davidson and Donny Channel, both from Westport, Washington, raced to complete the task.

"The forecast had called for seas to thirty feet," recalls John Magoteaux. "But the seas were worse than that. And the wind was howling! Just smoking!"

With his ship sinking steadily beneath him, Tom Miller raced across the back deck of the crab boat. The deck was slippery with water draining away from the last monstrous wave. He checked the stern hatch cover. It seemed secure. Then he sprinted back toward the safety of the bow-mounted wheelhouse. Just as he reached the door to the galley, another thirty-foot breaker crashed over the entire ship. The wave drenched him with spray and flooded the deck, pinning the galley door shut.

In an effort to escape the rising water, Miller scrambled up the superstructure of the wheelhouse and entered through the back door. Yet even after he had reached the temporary safety of the ship's bridge, he realized his reckless climb had been in vain—for his ship was still taking on water somewhere near the stern. Standing inside the wheelhouse, Miller and his three-man crew could only watch as the stern deck of the seventy-five-foot crab boat settled deeper into the churning black sea.

Then without warning, the main engine died, the steering system went out, and immediately the huge seas threw the ship sideways. Tossed by the waves and adrift in the spray-filled darkness, the ship continued to settle. Soon the storm waves pummeling the boat began breaking over its entire length and water began pouring in through the vent pipes and engine-room stacks mounted on top of the wheelhouse.

Wave after wave threatened to bury the vessel. Inside, in the gyrating space of the ship's galley, the freezer broke loose and sailed across the room. Below, in the engine room, tools and equipment and spare batteries tumbled pell-mell through the darkness.

In the wheelhouse, the crew was trying to send out a Mayday call for help when the boat rolled onto its side. Seawater soon drowned out the auxiliary engine, and the crew found themselves suspended in darkness, with no power by which to steer, see, or send a message.

With a flashlight in hand, Donny Channel made his way to the engine room, which was filled with chest-deep water. As he waded in, he spotted a can of starter fluid floating by. He grabbed it, gave the auxiliary engine's intake a shot of ether, and cranked it over. The engine coughed and started.

Then the lights flashed back on and Channel thought, Okay, you guys! Somebody better be sending out a Mayday, because this is it! It better be Mayday time up there in the wheelhouse!

Seconds later, the *Master Carl* rolled completely on its side and remained there. Once again, the water flooded out the auxiliary engine. Channel knew there was nothing more to be done, and in the liquid darkness, he scrambled back up into the wheelhouse and joined his crewmates in donning their survival suits.

"Tom Miller was really the hero of the thing," recalls John Magoteaux. "He got things organized and, one by one, got us out of there through the side window on the top of the house. Then we got the raft off over the side and he made sure each one of us got into the raft before him."

Once inside the raft, the crew members found that they were still tied to the sinking vessel. Each time the ship rolled, the keel flashed past them, the raft line came tight, and the shape of their raft contorted. They were in real danger of being pulled under, but no one could find a knife. Finally, in blind desperation, Magoteaux began chewing on the line in an effort to sever it with his teeth before the boat sank and dragged them all down with it.

"Then Tom remembered he had a pocketknife," recalls Magoteaux. "But he had to take his suit halfway down to get at it. The waves were crashing in on us, and with his suit still half down, we held on to him as he leaned out one end of the raft. Though he got repeatedly drenched, Tom managed to cut the line, or that would have been it for us! He got soaked, and with his suit half full of water, he never did get warmed up after that."

No sooner had they cut themselves free of their sinking ship than the typhoon-force winds whipped them off into the spray and darkness. Lifted and tossed by the heaving waves, and with the storm winds roaring constantly across the roof of the dome-covered raft, they could only drift and wait for daylight, then some seven hours away.

Most frightening of all, however, was the roar of unseen storm waves breaking as they approached from off in the darkness. Some waves collapsed directly down over them, flattening the raft and its occupants and tossing both about with an incomprehensible power and fury. Other waves passed beneath them, and then the crew of the *Master Carl* could feel themselves rising steeply, and falling sharply. It was like being blindfolded on a never-ending roller-coaster ride. "The worst part was not being able to tell which monstrous breaker had your name on it," recalls Magoteaux.

The night was only half gone when one incredible wave sent their raft tumbling down its face, pitch-poling, end over end. When the wave finally passed on, the crew found themselves standing on the raft's roof, submerged to their necks in the icy thirty-nine-degree Gulf of Alaska water.

Moments later, Donny Channel and John Magoteaux dived out from under the inverted raft. "It was the only thing we could do," recalls Magoteaux. "Somehow, we had to get that raft turned upright. You couldn't think about what it was like out there in the water in the darkness. We knew we had to get that raft upright if we were going to survive."

Channel and Magoteaux grabbed the outside of the raft and screamed instructions for those inside. They were to disperse their weight on the down-wind side of the raft while Channel and Magoteaux lifted an edge of the raft in the hope that by exposing the surface to the fierce gusts of wind, the raft would be blown back over. The plan worked perfectly, for when the wind struck the uplifted edge, it instantly flipped the raft back upright.

But just as quickly, those outside the raft found themselves bucked off the life-preserving form of their raft, which, propelled by the fierce winds, was streaking away from them and off into the night. In that instant, the terror and hopelessness of the moment struck them, and they faced the fact that they were caught in a fierce open-sea storm tens of miles from any shore, and that their raft had just taken off! It was moving far too fast. There was no way to catch it.

Then a length of line, the same line that once had held them to their sinking ship, came streaming past them. Channel and Magoteaux groped blindly in the darkness and somehow managed to grab it. Terrified, they pulled themselves back through the seas to the drifting raft and crawled inside through one end.

Back aboard the raft, they worried. They knew that the next time the raft flipped and failed to right itself, the chances of surviving were slim. They decided that the best possible way to stabilize the raft was to flood its floor with seawater. The low center of gravity produced by nearly a foot of water in the raft worked perfectly, and the equally vicious waves that followed failed to flip them.

Now, the four crewmen fought depression, seasickness, and defeatist attitudes. They tried to talk about positive things. They figured they'd hit Kayak Island, and if they missed its rocky shores, they knew they'd be blown into Prince William Sound. Maybe they'd end up on Montague Island, or be intercepted as they drifted along the way. There was no way to know whether anyone had heard their brief Mayday message, but since no one had acknowledged the call and no coordinates had been exchanged, those in the raft held out little hope.

The storm was one that fishermen from Kodiak to Cordova often would recall over the next decade. Late that night, the searing one-hundred-knot winds peaked, easing to about forty knots, but the waves remained huge.

With daybreak came the first sighting of land. Miserable with cold, those inside the raft could hear the building roar of the pounding surf. It was too misty and the seas were still far too rough to allow a clear sighting, but they could see the hazy gray form of an island. At the base of it, they spied what they believed was the flash of surf waves spending themselves on the beach. What they didn't know was that they had spotted a white strip of cruel breakers exploding across a rock reef, a reef that stood between them and the beach.

Completely at the mercy of the wind and seas, they were blown closer to shore, where seas began to stand up on them. Their raft rose steeply and fell sharply, and as they drifted, they moved through sparse fields of kelp, and heard the curious cries of sea gulls circling overhead. Then immense breakers began to power past them.

Screaming to be heard, the crew of the *Master Carl* soon agreed that it would be suicidal to go in through the surf inside such a contraption. One by one, then, they crawled outside the raft, clutched the nylon cord that rimmed it, and held on tight as the windblown raft streaked toward whatever lay ahead.

The sound of the breakers crashing on the reef grew proportionately louder as they drew closer. One of the crewmen didn't know how to swim, and the building roar of the surf terrified him.

Sensing something behind him, Magoteaux whirled in the water. A huge breaker loomed directly overhead. "Oh, my God!" he gasped. He had his arms double-wrapped in a death-lock around the raft's cord when that wave collapsed upon him. "It drove the raft and myself straight down," he remembers. "It tore me off and just sent me shooting through the water, straight down, probably fifteen or twenty feet. I couldn't tell which way was up. I was almost out of air when I forced myself to relax.

"Not knowing which way to swim, I let the buoyancy of my suit carry me back toward the surface. Once I saw which way I was drifting through the water, I started swimming in that direction for all I was worth—kicking, pulling, stroking.

"The water had been dark green below, and as I neared the surface, the water started getting lighter. Then I could see the surface. I popped up and looked around, but without my glasses I couldn't see too well."

Donny Channel didn't think anybody would make it. The size of the waves astonished him, but there was no turning back. They were committed.

That first wave caught everyone by surprise. It smashed Channel under and knocked the wind from him. Submerged and blinded by the foamy froth, he felt himself being dragged into the depths and tossed wildly about by the brutal explosions of collapsing sea. The currents continued to pummel him, and he felt icy rivulets of the Gulf of Alaska water pour in through the rim of his survival suit's hood.

Driven down several fathoms beneath the surface, Channel covered his nose and mouth, and, holding on to what little air he had left, tried to keep from blacking out. He, too, had decided to wait until his suit floated him back to the surface. He was determined not to inhale the cold ocean water, so he curled up in a protective ball and waited. It seemed like whole minutes passed before he finally bobbed to the surface.

Then he broke into the clear. He spit up water and coughed heavily, and sucked in deep lungfuls of the fresh cold air. He was trying to get his bearings when he turned and froze. For there, far overhead, in the very act of breaking, came the next wave, thirty feet and one hundred tons of sloping green water, roaring and falling in a thunderous, foaming tumult that collapsed down upon him.

Channel felt himself being driven toward the bottom, and although he somehow managed to keep from being crushed on the reef, he was close enough to know the gritty feel of sand between his teeth. He surfaced again, only to find himself pounded repeatedly by two- and three-story waves coming up from behind him. The roar and the limitless power of the storm waves were frightening and incomprehensible. Worse, he had no idea what type of shoreline lay ahead.

John Magoteaux had lost track of the others. He was clutching the rope encircling the life raft when "Donny Channel came popping up out of the water right beside me. He looked pretty bad, too. He'd swallowed a lot of water, and he had seaweed all over his head and shoulders. I grabbed on to him and got him up onto the raft. And together we were swept across the reef and into shore."

They staggered up on the beach and collapsed in the sand. When they could walk again, they began searching for their skipper and the other crewman. It was windy and cold, and they knew it was important that they remain in the warmth of their suits. However, bloated as they were with seawater, walking took an enormous amount of energy and they became exhausted after a short distance.

Magoteaux and Channel made the decision to lighten their loads. Unzipping their suits, they lay on their backs in the sand and raised their feet into the air. The seawater that had seeped into the suits now poured out over them, chilling them instantly. Resuming their search, they walked up and down the beach. Patches of snow lay on the high ground above them just inland from the sand. With a cold and blustery wind whipping constantly over them, the two men became severely chilled. Feeling leaden with fatigue, they soon were forced to give up their search.

Knowing that their own survival depended on finding protection, they pulled their life raft up off the beach, turned the raft upside down, and stuffed grass and wood around its upwind edge. Then, using it for a shelter and windbreak, they crawled underneath the raft and huddled together on the sand, where they immediately dropped off to sleep.

"Wake up! Wake up!" Magoteaux pleaded in an intense whisper. "And listen!"

Channel stirred awake. He could hear strange snuffling and grunting sounds coming from outside the raft. Curious, he lifted up one end of the raft and came eye-to-eye with an adult grizzly bear. The animal was so close, Channel could have stuck a five-gallon bucket on its head.

The two crewmen staggered to their feet and threw off the raft. The grizzly moved off about thirty feet and turned and faced them. The bear's nostrils drew in and out as it repeatedly sniffed the air. Its nose looked as big around as a man's thigh. Brisk gusts of wind moved in waves through its thick fur pelt.

Grabbing any rock that was within reach, the two men began pelting the unwelcome creature. "Get the hell out of here!" they shouted.

The grizzly stood on its hind feet then, sniffing the air and clicking its teeth.

At that moment, it struck Donny Channel how he and his crewmate must look to the grizzly. Clad in suits of rubber and neoprene, he thought: We must look just like a couple of sea lions, one of that breed of bear's favorite meals.

Perhaps the grizzly already had eaten, or perhaps it had never seen sea lions act so strangely, for the bear turned and wandered back up the beach. In spite of the unnerving confrontation, fatigue settled back over the pair, and they crawled back under the raft again and went back to sleep.

That night, the bear returned and began circling the raft, growling and tearing up the turf on all sides. Inside the raft, only inches away, Channel and Magoteaux could only wait and listen. They had one rocket flare ready and waiting to use if absolutely necessary. They planned to pull the pin and discharge the flare into the face of the bear if it actually came for them. But, although it was close enough for them to catch the scent of the creature's rank smell, not once did the bear touch the life raft itself.

By daybreak, the bear was gone, and the two crewmen abandoned their raft and began hiking down the beach in the direction of the lighthouse at Cape St. Elias. Fearful of being attacked, they reacted to each brown formation of log and rock by asking one another, "Hey! What's that?"

As they made their way toward the lighthouse, icy thirty-knot winds whipped constantly over them. John Magoteaux started showing the classic symptoms of hypothermia. "I couldn't figure things out," he recalls. "I ran out of energy. I couldn't go any farther. My body temperature had gone down to the point where I was almost hallucinating. I could barely walk. I couldn't think.

"Donny took care of me. He found a sheltered spot and took one of the wet hand-held flares and managed, somehow, to get that thing going. He built a fire out of plastic net and buoys he found on the beach. I tried to help but managed to find only two wet sticks of driftwood. I was slowly starting to lose my coherence. I don't know how Donny did it. I owe him my life."

That afternoon, above the sounds of the pounding surf and blowing wind, the two crewmen heard what they at first thought was a plane making its way across the island. As it drew closer, Channel could tell it was a helicopter.

Immediately, he sprinted over and grabbed the only remaining flare and tried frantically to get hold of the small metal ring that would light it. But his fingers were cold, and as his hands fumbled after the ring, his eyes flashed between the flare ("to see if I was doing it right") and the horizon in an effort to spot the helicopter. By the sound of it, the chopper had to appear at any moment.

The clacking roar grew louder with each passing second. The distinctive white and orange paint and the black bumblebee nose marked it as belonging to the U.S. Coast Guard. It was moving fast along the shoreline just off the beach. Only seconds before, it had flown past their raft without spotting it. It would soon be past them, too, if Channel couldn't get the single remaining flare ignited in time.

The helicopter was almost overhead when the flare went off. For a moment, Channel thought the rocket was going to strike the chopper's front windshield. Then the helicopter's engine growled, its tail rudder dipped, and its black nose rocked skyward. It seemed to Channel that the helicopter pilot stopped in almost no distance at all.

"All right!" screamed Channel and Magoteaux repeatedly as they stumbled forward.

Once aboard the helicopter, their joy at being rescued was short-lived when, only a short distance up the beach, the copter pilot came upon the bodies of the heroic young skipper Tom Miller and Tom Davidson. They had apparently been killed while coming in through the surf.

PART
ONE

THE GREENHORN
SEASON:
TANNER CRAB FISHING
ABOARD THE *ROYAL
QUARRY*

ONE

As I pounded the waterfront of Kodiak's Cannery Row in search of work, I stretched my black wool cap down over my numb ears and withdrew farther into the warm, protective folds of my Navy pea jacket. Broke and unemployed, with virtually no experience at sea, my stomach felt empty with apprehension. Huddling against the cold, indifferent winds of January, waiting, literally, for my ship to come in, I tried to imagine what Donny Channel and John Magoteaux, as well as the two men who died, Tom Miller and Tom Davidson, had endured.

Several days earlier, a hard-drinking fisherman in the B & B Bar had told me the *Master Carl* story. For a bar-stool rendition, it was surprisingly accurate, except that he had the grizzlies chasing Magoteaux and Channel for two whole days. "And the bears would of got 'em, too," he put in finally, "if the Coast Guard wouldn't have come along and chased 'em off!"

I'd arrived in Kodiak with twenty dollars to my name, but I held one ace card. I had the address and phone number of crab-boat skipper Mike Jones. He was said to be a young, heads-up fisherman who owned and ran the seventy-one-foot blue and white steel crab boat, the *Royal Quarry*. We'd gone to the same college (Oregon State University in Corvallis) together.

"I can't promise you a thing if you come up, Spike," Mike Jones (Jonesy) had told me when I phoned him from Oregon only a week before. "But this I *can* guarantee you. You're not going to get a fishing job sitting down there in Portland. This is where the work is. There's some real money to be made up here. Right now, my men are pocketing about a thousand dollars a week. And that's fishing tanner crab [a tan-colored spidery crustacean that grows to several

3

feet in width, averages about three pounds in weight, and is marketed as snow crab]. The real money is made during king crab season. You wouldn't believe how much you can make, provided you get a job on the right boat.

"But it's tough work. Dangerous work. Dozens of crewmen get killed every year up here. And not just anybody can do it. The crab pots are called 'seven-bys' [seven feet tall, by seven feet wide, by a yard deep] and they weigh seven hundred and fifty pounds completely empty. When they arrive on deck, they may weigh a ton if they're full of crab. You've got to be able to control their swing enough to guide them into the rack [pot launcher]. And you need to be able to do it in rough seas. The weather can really get tough up here.

"If the fishing's good, you might have to work steady for thirty, forty, or fifty hours without sleep, or even a cooked meal. We can teach you enough about knots and navigation to get by, but the one thing that no one can teach you is how to work. You can't wait to be *told* what to do. You've got to be able to *see* it, and not be afraid to jump right in and get it done. But you've worked as a logger. You probably already know what that's about."

He paused.

"If you're really serious about coming up, I can find you a place to crash for a few days," he said finally.

Three days later, I flew into the Kodiak Island Airport.

With a one-way ticket in hand, I had caught a flight out of Portland, Oregon. Some fifteen hundred air miles later, my plane had touched down in Anchorage, where I climbed aboard an Alaskan puddle jumper. My journey took me south along Cook Inlet and out across some 270 miles of Gulf of Alaska waters to Kodiak.

It was a sleet-filled night and my plane was late getting in, but Jonesy was there to greet me. He gave me a pickup truck ride into town and left me with one of his crew members, Steve Calhoun, a good-natured young man from northern California. Steve was exhausted from a long day of repair work aboard the *Royal Quarry* and after only a few minutes of conversation, he turned in.

I was fast asleep in a sleeping bag on his couch when, long before first light, Steve rose and dressed in boots and work clothes. As he walked out the door, he paused. "Here's my key," he said, setting it on the kitchen table. "You can stay here while we're gone. Just be sure and latch the windows and turn everything off and lock up whenever you go out. We should be back in about a week."

With a long green duffel bag in hand, he opened the door to go. It was dark outside. Several inches of fresh powdered snow had blown in across the door-step. "We've got grocery supplies to store away and a few chores to get done on the *Quarry* this morning before we leave," he continued. "So, we shouldn't be shoving off until afternoon. When you get up, come on down and I'll show you around the boat."

"Thanks a lot," I said, my voice filled with gratitude.

"Oh, hell, that's all right, ah . . ." He groped for my name.

"Spike," I offered.

"Oh, yah, Spike. Anyway, I don't have any trouble at all remembering what

4

it was like to be broke and trying to hunt up a job on one of these crab boats. No trouble at all."

"Good fishing to you," I replied.

Then, with a short wave of his hand, he lifted his duffel bag and fled out the door. I shuddered as the cold blast of arctic air he had let in rolled across the floor and over me. I thought of the brutal weather, my inexperience at sea, and the sad state of my personal finances, and suddenly a moment of panic and doubt gripped me. I asked myself what could have driven me to take such a risk; to leave family and friends and all the warm familiarities of home to venture into this primitive northern country and compete for a crewman's berth in such a brutal and deadly trade.

I slept hard and long, and when I rose again, the morning was nearly gone. I chose to skip breakfast, dressed hurriedly but warmly, and headed straight for the waterfront and the B & B Cannery in an effort to catch the *Royal Quarry* and her crew before she shoved off.

Shelikof Boulevard, a street of mud and gravel, wound its way from the center of town along the waterfront to an area commonly referred to as Cannery Row, a strip of processing plants standing shoulder-to-shoulder at the edge of St. Paul Harbor.

With the season already in full swing, Cannery Row was already alive with activity. Everywhere the thriving capitalistic pulses of movement and noise and commerce prevailed. All along the waterfront, one could hear the shrill tooting of the shift whistles, the deep belching sounds of crab boats as their powerful diesel engines growled to life, the hydraulic groan of straining dock cranes, the hum of electricity, and the metallic click of chain links meshing over conveyer-belt sprockets. And there was the hiss of escaping cooker steam, the revving of motors, and the warning beeps of racing cannery hysters (forklifts), while out in the harbor, sea gulls squabbled over discarded scraps, and dockside, cannery workers joked in Taiwanese and Filipino tongues.

To the nose of the pragmatist visiting Cannery Row drifted grim waves of lung-stopping ammonia, the rotting smell of oozing cannery slime, and the unrelenting odors of iodine, kelp, and decomposing microorganisms. Even though I sported an empty belly, those smells renewed my spirit, filled me with a sense of opportunity and adventure, and sent romantic notions jolting through me.

There was an energy here; a boomtown intoxication that can be found nowhere else on earth. There was a sense of optimism and imminent excitement all around; life had taken on a new meaning in these parts. One could see it in the shuffle of the crab-encumbered ships vying for off-loading space along the waterfront, and in the passing faces, and in the quick, restless, straight-ahead way people moved. I could sense an excitement in the tired but grateful faces of the cannery workers as they shuffled home from another profitable sixteen-hour shift. And in the grocery stores, where rich and hungry crab-boat crews bought up every loaf of bread, gallon of milk, and sirloin steak in town, regardless of the price.

5

I could see it in the crowded streets, where long bumper-to-bumper rows of work vehicles filled every available space and along which dual-wheeled trucks with precarious loads of 750-pound crab pots balanced on their flatbed trailers raced. They were on their way to be unloaded at the earliest possible moment onto their ships, which were tied to the tall wooden pilings dockside.

Their thudding tires pounded a track through the rough, muck-filled chuckholes and launched geysers of ash-based mud out both sides, coating the long rows of parked vehicles from hubcap to door handle with a sloppy goo the color of wet cement. Jockeying for position on the docks, these same pot-toting trucks had hardly rolled to a standstill before their crews clambered atop them. The young men who leapt to untie the pots moved with excitement and vigor. And as I watched, I found myself envying them, the positions they held and the lives they lived, for they had become an integral part of something vital and adventurous.

I'd been part of similar scenes before. Though now still several years shy of thirty, I was there when ambitious men, driven by soaring oil prices, rushed to open up new oil fields outside Morgan City, Louisiana; and I was there when soaring lumber prices lured the same kind of men to plunge, axes in hand, into some of the last great stands of prime old-growth timber in the United States, near Forks, Washington, on the Olympic Peninsula.

As I wandered broke and jobless along the waterfront, I tried to remind myself that I had not ventured north completely untrained. While working in the offshore oil fields bordering the Gulf of Mexico, I had served out an apprenticeship as a deep-sea diver; and during that tenure, I had learned to cut and weld underwater, and how to inspect ship hulls, barge hulls, and propellers.

While logging in some of the more remote camps in Washington, Oregon, Idaho, and southeast Alaska, I'd learned how to work around heavy equipment, how to stay out of the bite of a straining line, and a good deal of the common sense of serious labor.

I'd learned that a man could find a niche for himself in life just about anywhere, if he kept his nose clean, his mouth shut, and worked without quarter or need of praise.

Call it impetuousness, or point to the insatiable greed of man, but here in the remote Alaskan outpost of Kodiak, simple American capitalism once again had touched off a modern-day gold rush. Yet this time, the prize was not oil or timber. This time, the nuggets were showing in the tanks of fishing boats returning to port to be off-loaded. During the winter months and into the spring, their cargo came in the form of tanner crab, and during the fall, as spiny red king crab.

Experienced skippers—their giant steel ships ranging from 70 to 170 feet—with the knack for tracking down and boating the milling, wandering, unpredictable "bugs" were said to be grossing as much as one, and, occasionally, even two million dollars a year! And the crewmen working on those elite boats were earning upward of 7 percent of that gross, or $70,000 to $140,000 per man!

Those who had caught the scent of such vast riches let nothing deter

6

them—not the gamble of a plane ticket north, the cold inconvenience of winter, the fear of the unknown, or harrowing tales of shipwrecks and disaster. Now, hundreds of would-be fishermen were pouring into the port villages of Kodiak in the Gulf of Alaska and Dutch Harbor in the Aleutian Islands on the edge of the Bering Sea. I was no different.

I arrived at the B & B Cannery just in time to wave goodbye to Jonesy as he shoved off from the docks.

Jonesy saw my hurried approach on the dock above him, and poked his head out the wheelhouse window and yelled up at me. "We'll be gone about a week, depending upon the fishing," he called, the white jets of his breath lingering in the cold winter air. "Check with Joe at the cannery to see when we'll be getting back into port. We keep radio contact with them. They'll know when we're scheduled to be back in port."

"Who?" I yelled after him as his ship pulled away.

Jonesy again stuck his head out the side wheelhouse window, pointed to the cannery building behind me, and hollered back, "Joe! Hungarian Joe!"

Standing on the edge of the cannery dock, I stuffed my hands into the warm wool pockets of my pea jacket and watched the *Royal Quarry* idle out of the harbor. I was still there when the toylike spectacle of the blue and white ship turned the corner and moved out across the wrinkled gray expanse of Chiniak Bay.

Suddenly, the commotion of the waterfront broke into my thoughts.

"Hey, Joe! Where do you want the pallets stacked?" a youngster in bright yellow rain-gear pants called from the steamy entrance to the cannery.

I knew it was Hungarian Joe standing behind me even before I'd met him. He'd been part of the Hungarian Revolution of 1956, I'd been told. When several divisions of Russian tanks rolled into his country, he'd escaped into Austria. Dressed in plain greens and browns, he wore a flat-billed baseball cap and answered the rain gear–clad youngster with a heavy, Eastern European accent.

"Joe," I interrupted, "I was wondering if you know how often the *Royal Quarry* reports in?"

He did not seem to hear me. "Vee have crab to cook now!" he replied. Then he pivoted and hurried back inside the cannery. He wasn't ready to talk with anyone then. I'd catch him first thing the next morning.

I turned back to the water and once again took in the view. On my far right, and slightly inland, I could see Mt. Barometer, a unique and pointed mountain that rose abruptly for some three thousand feet. It stood unchallenged, and greeted all incoming flights at the end of the airport runway.

In the distance stood the long, rugged shoreline of Chiniak Bay. The low-lying country was covered with weather-stunted spruce trees, and its headlands played out abruptly in the form of cliffs at its waters' edge. Inland rose an entire range of steep round-topped mountains with broad bowl-shaped valleys stretching between them. Buried in snow, with bare black rock showing through in places, their treeless slopes offered a bleak reminder that winter was full upon the land.

7

Behind the gigantic hills near the water stood a range of four-thousand-foot glacier-covered mountains, their tallest peaks disappearing into the clouds. But they looked much taller than that, rising from sea level as they did. As I watched, I noticed how the wind was blowing a haze of powdered snow up and over the mountain ridges. Blown into feathery wisps, it looked like mist rising from a breaking surf.

Ahead of me, across the channel, lay Gull Island, a narrow strip of land—a plateau of rock, actually. Crowned with brown winter grass, raked by an icy wind, and encircled by a lonely gray expanse of sea, the island was accented with a tide wash of rich brown kelp and the plump white-breasted forms of sea gulls taking refuge along the rocky shore.

Not twenty feet from the blinding arc light of a welder cursing his trade, I watched a narrow rainbow-colored strip of spilled fuel oil drift by. Submerged in the depths below it, white-stalked sea anemones as long as one's arm sifted the rich currents. Then there was a submarine flash of fin and blubber, and seconds later, perhaps a hundred feet off, the marble-eyed, mustached head of a sea lion rose from the water. Rich brown in color, its fur wet and slickened, the frolicking creature barked, rolled, then dove again, vanishing behind the cloud of its own breath.

Fatigued from the cold and the travel, that night I hiked back to Calhoun's apartment and crawled into my sleeping bag. I felt thankful to be out of the weather and slept that night like the living dead.

The next morning, I rose before sunup, downed two glasses of whole milk, four pieces of wheat toast, a quarter pound of bacon, and a half dozen eggs, then once again made straight for the waterfront. I still had hopes of stumbling upon a skipper who, like Mike Jones, knew how to find and catch crab, treated his crew in a straightforward manner, and had a reputation for keeping the ships he skippered afloat. It was a high and mighty order, I chided myself, for someone with no experience at sea and with little more than a sawbuck left to his name.

Down on the city docks, I came upon what I believed to be my first crab boat. I placed both my hands on the steel side railing and cleared it in a single motion. Then, standing on her back deck, I came face-to-face with a sign posted on the wheelhouse door. It read FULL CREW. Farther down the docks, I came upon still another. It read DON'T EVEN ASK!

Then I checked Cannery Row for new arrivals and spied the fishing vessel *Cape Fairwell* tied to the pilings in front of the B & B Cannery. "Do you guys have a full crew?" I asked, calling down in a subdued voice to a crewman working on the back deck of the crab boat. Heavily clad, the young man appeared to be splicing together two lengths of yellow cord line.

Without looking up from his labor, he shook his head. "You're the third person to ask me that so far, and we only got back in port last night." Then he glanced up at me. "But that doesn't mean a thing," he put in. "You gotta just keep on asking. Guys get injured. Guys get fired or move on. The worst a skipper can say is no." Still, being refused struck me as the deepest form of rejection. And when the crewman returned to his splicing, I silently moved on.

8

For an entire week, I pounded the waterfront in search of work. Each day, I made it a point to stop in at the B & B Cannery and check with Hungarian Joe as to the possible arrival of the *Royal Quarry*. Then, with the daylight gone and my ambition waning, I'd drift from bar to bar, keeping my eyes peeled and my ears open.

The bar life in boomtown Kodiak was the basic social activity for those crewmen in port. It was an all-night lifestyle in which spirits were downed and adolescence could be prolonged indefinitely. It was a stopgap for the missing loved ones of home, and deckhands readily substituted the blood-warming oblivion of liquor and the wild bohemian camaraderie in their place.

Many of the bars reminded me of museums in which drinks were served. King crab and tanner crab that had been injected with formaldehyde were commonly mounted over the bar; along the walls hung pictures of crab boats, harpoons, nets, buoys, glass balls, and life rings. The legal drinking age was nineteen. Some of the bars didn't even blow reveille until 5:30 A.M. And in the long daylight hours of late spring and early summer, I was told, you walked out the front door blinking into daylight.

There was Solly's Bar (and restaurant), Tony's, The Mecca, The Ships, Shelikof Lodge, The Harvester, The Anchor, The B & B, and the Beachcomber. And I came to know them all.

Shying at first from the drunken commotions I chanced upon, I remained silent and waited for the others to do the talking. Each night, I listened to unrefined men clad in muddy boots, wool caps, work-worn jeans, and heavily insulated jackets speaking in a lingo all their own. They talked with enthusiasm about strange (and therefore exciting) places I had never been: Whale Pass, the "south end," Portluck Banks, the bays of Kajulik and Chiganaga, Afognak Island, Cape Alitak, the Barrens, Homer, Icy Cape, Shelikof Strait, the Semidis, Lazy Bay, Dutch Harbor, compass rose, the "slime banks," and "the mainland." These exotic names, in turn, were tied to anecdotes involving catch totals and crew earnings, stormy seas, and men who had experienced them.

These men spoke a nautical language, one of geography and events gleaned from decades of open-sea experience. I learned of the fierce tides of Sequim Pass, the prop-grinding shallows of False Pass, the dog-salmon runs of Kukak Bay, the deer hunting on Kodiak Island (with a legal limit of five deer per person per season), the elk hunting on Afognak, the bleak loneliness of the Pribilofs, the mind-bending savagery of seasons "served" around Adak (out in the Aleutians), and of adventure in the uncharted waters along the Alaska Peninsula.

So I eavesdropped without shame, and by week's end began to probe with questions of my own. I bought drinks for complete strangers and asked questions until further questioning would have seemed rude, or neurotic.

One night, I stepped out of a bitter cold and into a wild and spirited event at a small, waterfront bar. It was there that I first saw her. She was a good-looking brunette of about twenty or so, a young spitfire of a woman. At the moment, she was locked in an arm-wrestling battle with a boisterous greenhorn who had just

9

flown into town. Unable to ignore his mouth any longer, the young woman had offered up a small challenge. Now an intense throng of hard-drinking fishermen and cannery workers were pressing forward on all sides to get a view of the match, and to voice their support.

The commotion was deafening. A rough-looking fisherman sporting a foot of beard and a gruff voice was cheering her on. "Show him what you're made of, Susey!" he growled, pounding his fist on the bar.

One fanatical female booster screamed, "Come on, Susey! Take him! Pin the bastard! You can do it, Susey!" As the woman cheered, she pounded on the back of a fisherman with one hand, while slopping beer on the floor from a glass held in the other. And as the battle at hand continued, and the young girl named Susey drew closer to victory, she leapt and screamed and her voice climbed into an hysterical shriek.

One had only to look into the greenhorn's face to predict the outcome. And as the amazing vein-popping muscle of Susey's right forearm drove his arm closer to the dark hardwood surface of the bar, his eyes looked bloodshot and his face had flushed purple with effort. Then, with a final gush of exhaled breath, the greenhorn's arm collapsed in surrender and the room erupted. The embarrassed greenhorn fled from the friendly jeering and pats on the back. With his head down, he moved hurriedly through the crowd and out the door.

As I departed, the attractive young woman called Susey, or, more affectionately, Susey Q, was sitting at the bar with a warm glow on her face. She was surrounded by a number of admiring young women. Then with a gentle, feminine flair, she swept the clean brown hair from the side of her face and, with a calloused hand, daintily lifted a fresh cold glass of beer to her lips.

Later that night, I was listening to the music and jamming down straight shots of tequila at the Beachcomber Bar when *Royal Quarry* skipper Mike Jones appeared. He stole one of my six shot glasses of tequila and swilled it down before I could protest.

"Jonesy, you're back!" I shouted, feeling surprised and delighted to see him. "When did you get back into port? They told me you wouldn't be in until tomorrow! How did your trip go?"

"Well, we've got nearly a full load on board," he replied matter-of-factly. "But we were lucky. We didn't break down, and the weather and fishing were good."

"You've been 'lucky' all season long," I joked, poking fun at his modesty. "In fact, I heard your December king crab season wasn't exactly a bust, either."

Jonesy broke into a wry smile. "How did your job hunting go?" he asked.

I shook my head. "Not nearly as well as your fishing trip," I said.

Then the skipper of the *Royal Quarry* turned to me in earnest. "Spike, we've got a man on board who just might not work out. Now, mind you, I can't promise you anything in the way of a job, but if you think you'd like to, well, you could go out with us this next trip. It'd be a way for you to learn the ropes and see what it's all about and how you like it. And it would give us a chance to see how you work out."

I could feel my hopes soar.

"Now, I can't pay you anything," he continued, doing his best to be clear about the terms. "And I can't promise you a job, because I have a full crew on my back deck right now. But they're an unhappy bunch at the moment. Ordinarily, that wouldn't bother me; I mean if they were mad at me, that wouldn't worry me much at all. But the problem is, one crewman doesn't seem to be able to pull his full share of the load. That makes for bad feelings on deck. And I don't like to see that.

"Besides, if things went well during your trip out with us . . . well . . . I can tell you it wouldn't hurt your chances at getting aboard our ship, if and when it does come open. And if things don't work out in the way of a job on board our ship, once you get a little experience at sea, opportunities have a way of opening up for a guy who keeps trying. But you have to be patient. And sometimes you have to pay a few dues first. You have to use your head."

"Jonesy," I replied, my eyes wide at the opportunity, "I'll be ready whenever you are. You can damned well be sure I want to go! You betcha!"

He seemed pleased at my enthusiasm. "I thought you might," he said, smiling again and swilling down the last of his drink. "Steve Calhoun will be buying supplies for us tomorrow morning. Tag along and help him if he needs it."

He rose to go. "You'll need boots, gloves, and rain gear. If you've got any questions about what to buy, talk with Calhoun. He'll set you straight."

Early the next morning, I bumped into Hungarian Joe at the B & B Cannery. "Soooo," he began, "did you hear of zee load Mike Jones brought to us?"

I nodded.

"So," he began again, "you go out with him on next trip. Yes? No?"

When I nodded a second time, Joe's eyes danced with delight for my good fortune.

"But I'm not getting paid," I put in quickly. "This trip is just a test to let me see what it's all about . . . and to see if I work out."

"Ah, you vill make it!" shot back Joe. "You must vait and see!"

I nodded again. I always liked people who genuinely wished the very best to others.

"Ahhh, this Jonesy, he know how to find zee crab," he continued. "He one damned good fisherman. He deliver only to us. He come to port and bring all zee ship vill carry!"

I felt encouraged as I climbed down the dockside ladder and leapt down on the deck of the *Royal Quarry*. Steve Calhoun had arrived just ahead of me, and came out to welcome me aboard. We needed to buy groceries first, then we'd stop by Sutlifs Hardware Store and buy some working clothes and rain gear.

"What size boots do you wear?" he asked. "I have a spare pair you might be able to wear."

"Size thirteen and a half," I said.

Calhoun looked at my feet and made a face. "Damn, man," he said. Then added, "Well, so much for that idea!"

Suddenly, a young woman's voice broke through our conversation. "Hey!

11

Let's have a little less *talk* and a whole lot more *work* going on out there!" she hollered. Then she stuck her head out the back of the wheelhouse door and smiled.

I froze in my tracks. My mouth fell open. It was the girl who'd won the arm-wrestling match the night before.

"That's Susey," offered Calhoun, smiling at my look of astonishment. "She's your new deck boss."

TWO

On that cold and stormy February night, it was near midnight when Jonesy leaned out the starboard wheelhouse window. "Cut us loose," he ordered. "We're gone."

Susey and Calhoun ran to throw off the hawser lines and our crew of five drifted away from the protection of the dock and out into the wind. With the majority of the fleet out on the fishing grounds, the waterfront sat empty and quiet. We idled ahead at "no wake" speed through the sharp edge of a harbor chop and slid past the tall wooden pilings and blocky aluminum buildings of Cannery Row.

Once past Cannery Row, we accelerated ahead, winding our way through the buoy markers and past the Sea Land loading docks (where refrigerated truck vans full of frozen seafood were lifted aboard freighters for the trip south), and passed into darkness beyond. And with flickering long rows of the U.S. Coast Guard airport runway lights heaving in the darkness off our stern, we left Kodiak in our wake.

I hadn't said anything about working with a woman on deck, but secretly I cursed my luck. With several hundred crab boats in the Gulf of Alaska and Bering Sea fleets, what were the odds, I asked myself, of working on deck with the only female deckhand in all of Alaska? I worried that if she had trouble pulling her fair load, I and the others on board would have to take up the slack. But I decided to let the difficult tasks ahead do the sorting. Long ago I had learned that hard work had its own way of thinning the ranks. Those who could hack it stayed on; those who couldn't moved on.

The waters of Chiniak Bay were steep and building, and we'd been gone less than an hour when the first cruel pangs of seasickness hit me. My stateroom was perhaps half the size of the average urban bathroom, and as I huddled in the reeling darkness, fighting to hold down my dinner, I could smell the light, nauseous odor of herring and the acrid smell of diesel fumes wafting in the close air.

We were forced to drive headlong into the southeasterly seas, and as we soared over each new wave, my stomach and its contents floated free, only to come crashing back down in perfect cadence with our plunging bow. We were still making our way out of the harbor when I leapt from the warm, dry comfort of my bunk and fled in my stocking feet through the galley and out into the cold fresh air of the back deck.

I felt deathly sick—wretched beyond anything I had ever known. I bent and began purging my stomach of all its misery. Then, in an eerie lapse, the sounds of the seawater slapping against our hull and rushing past me all around hushed, and out of the corner of my eye I glimpsed the closing wave. Suddenly, I realized that I had wandered too far from the protective steel wall of the wheelhouse. In the next instant, a full fathom of bone-chilling, heart-stopping, Gulf of Alaska sea knocked me as flat as a flounder and swept me halfway across the deck.

It was a crude and unceremonious baptism, and as I gasped for air, all the youthful pride and naïve romanticism I had harbored about the life of an Alaskan crab fisherman vanished in that moment's cold wash. Instinctively, I filled my lungs to protest such inhuman treatment. It would have been comforting to tell someone and know that they cared, but then shame filled me. It had been a foolish error to venture out so far, and I stifled my screams and danced backward to the shelter of the wheelhouse wall and awning. As I retreated, my drenched cotton socks trailed nearly a foot beyond the ends of my toes, stretched there by the pull of the collapsing sea and wild, foaming deck wash. God, how I hated the moment, my life, and the blind ambition that had carried me here. Then, drenched and shivering, with my knees bent and legs braced wide against the twisting deck, I stooped and threw up for ten uninterrupted minutes.

The deck bucked and rolled like a cruel beast as I made my way back inside. Soaked, dripping seawater from every limb, I was drying my head with a towel when Susey came into the galley. "What the hell happened to you?" she asked, taking a robust bite out of a roast beef sandwich.

"You don't want to know," I replied, catching a whiff of microwaved mayonnaise and gagging back the rising nausea. "And I'll be damned if I'm going to tell you."

She chuckled, sat down at the galley table, and paused to take another bite. "You picked a real beaut of a time to make a trip with us," she said. "The weather report says we're in for some mean weather. They're calling for seventy-knot winds to hit Kodiak tomorrow, with gusts approaching a hundred knots up in the Barrens. And the seas are going to be rough, too, from fifteen to eighteen feet! It feels like it's already getting bad out here.

"Once we get around Chiniak Head, we'll have it on our stern for most of the way down the length of the island. Then we'll shoot through Geese Channel and anchor up in Lazy Bay and wait for the storm to pass. Wasn't supposed to hit till tomorrow, though."

Susey took another bite of her sandwich. Thick globules of steamy mayonnaise oozed out its sides. I cupped my mouth and staggered toward my room. Once inside, I collapsed into my bunk, stuffed a towel in my mouth, and, for the next ten hours, gagged on the wet heave of nausea.

Only twenty years old, Susey Wagner was an attractive 135-pound brunette. She'd grown up in the Santa Cruz Mountains of California and had never been more than an hour from home when she first decided to escape north to Alaska. Eighteen at the time, Susey had grown tired of the Cloud-9 truck stop clientele, where, while still in high school, she earned minimum wages cooking and serving up food, beer, and wine.

From the beginning, she knew she would never be satisfied with just existing or only getting by. It was important to her to get somewhere, to achieve something and do it on her own.

The day Susey received her high school diploma, she moved out of her parents' house and away from her five brothers and sisters. She supported herself with a $3.25-an-hour job, while on the side she hired out as a baby-sitter and sold her own handmade pottery at flea markets.

Only days after arriving in Kodiak, Susey found a job in the local Skookum Chief Cannery. When she discovered that the cannery would allow you to work as many hours as you could remain standing on your feet, she thought she had died and gone to heaven. The opportunity to get ahead was everywhere, and Susey was soon taking home three hundred dollars a week.

For an entire year, Susey bunked at the Skookum Chief Cannery. She began, as did every new cannery worker, by butchering and packing crab on the processing line. But she was interested in taking a more physically demanding role and soon got assigned to off-load crab boats. It didn't take Susey long to realize that the plum off-loading job was to run the crane mounted on top of the dock. Seated inside in the dry protection of an enclosed cabin, its operator lowered and raised brailers (baskets) full of crab from the holds of crab boats tied to the Skookum Chief docks. Susey was determined to win that coveted job, but when she approached the dock foreman, he flatly refused. Later, when she went off shift, he made it clear to everyone that he didn't want any *woman* running that crane.

Susey was stubborn, and she bided her time, becoming friends with the main crane operator. Often, after she had finished pitching a brailerful of crab, while her male coworkers took their usual well-earned break, Susey would climb out of the boat's hold and scale the thirty feet of ladders to the dock up top for a chance to run the crane, if only for a few moments. If the foreman was nowhere near, the crane operator often would give her his seat and stand beside her and give her lessons.

15

When the crab-stuffed brailer had been hoisted up onto the cannery dock, Susey would lower another empty brailer into the ship hold far below. She would then race back down the dockside ladder, swing down into the crab hold, and begin tossing crab into the next brailer as fast as her hands could fly. Through such tireless persistence, extra effort, and hustle, Susey gradually learned to operate the crane.

Then one windy, rain-washed night, the crane operator broke his ankle, and no one but Susey knew how to run the crane. The cannery foreman had left specific orders that Susey not be allowed to run the crane. He "didn't want any foul-ups!" But he was gone at the time, and there was an impatient skipper stalking the docks with 65,000 pounds of tanner crab growing weaker in his live holding tanks with each passing hour. So Susey climbed up into the cabin of the crane, sat herself down in front of the controls, and went to work.

She was still hard at work off-loading the ship when the cannery foreman reappeared. It was a cold and rainy night, and the windows of the crane's cabin were fogged up—with the exception of the rag-wiped hole through which Susey watched and made her calculations—and the foreman didn't notice her. In fact, the cannery foreman didn't even learn of the crane operator's injury for some time. When he did, he inquired as to who, exactly, *was* running the crane.

"What?" he cried when he discovered that his orders had been ignored.

"But I wouldn't worry," one cannery worker offered. "She's doing fine! Hell, she's been running it for the past five hours! She's nearly done now. She's off-loaded the whole ship!"

The cannery foreman grew quiet and said nothing, but from that day on, Susey ran the crane.

Soon, however, Susey's attention turned toward the excitement—and money—of going to sea, and she set herself a new goal, one that had never been accomplished before by a woman: to win for herself a berth aboard a ship in the king crab fleet.

In the beginning, she freely admits, she lucked upon a fine teacher—her future boyfriend and a longtime fisherman, Danny O'Malley. It was O'Malley who first opened the way for her to go to sea. He worked on the ninety-foot crab boat *Atlantico* (skippered by Bill Jacobson). When the opportunity for her to go along on a free trip presented itself, she jumped at the chance.

Once at sea, Susey watched carefully, absorbing the crew's every move. She noticed how they manhandled the huge 750-pound crab pots, but soon realized that the work required more than brute strength. It also involved stamina, positioning, and technique, things Susey thought she could manage in time. The bulk of the work on deck involved coiling lines, tying knots, pitching crabs, chopping bait, and filling bait jars—as well as getting along with those around her.

Susey also paid close attention to the deck lingo and sometimes felt she was learning a foreign language. When the *Atlantico* docked in Kodiak at the end of that first journey to sea, skipper Bill Jacobson called down from the wheelhouse for her to secure the hawser line to the cannery piling. Susey could only shrug her shoulders. She honestly had no idea what the man was talking about. What was a hawser? And what was piling?

But Susey soon learned, and the next time the *Atlantico* tied up dockside, the skipper didn't have to say anything.

Susey went to sea for an entire month without pay—just for the experience. She worked outside on deck right along with the rest of the crewmen. Afterward, back inside the wheelhouse, she cleaned the galley, washed the cupboards, and swept and scrubbed the floors. She learned early on that when a good deckhand saw something to do, he or she went to it and didn't walk on and leave it for the next guy to do.

She worked hard and tried not to complain, and by the end of the month, she had improved dramatically on deck. She intentionally avoided the bull-like job of pushing pots, but in all other areas she worked right along with the rest of the crewmen. Life at sea was new and intriguing, but it was more than that, those who were with her in the beginning will tell you.

Watchful of Susey throughout her apprenticeship, *Atlantico* skipper Bill Jacobson recalls that she was "genuinely curious" about the sea around her. She took naturally to the knots, the navigation, and the endless work. She maintained a good attitude and managed to blend in with the men. Somehow, Susey fit.

Then one day, she remembers, an *Atlantico* crewman "slipped on the dock, of all places," and injured himself. Bill Jacobson took on Susey as the man's temporary replacement—at 4 percent of the crab boat's gross, a half share.

As one of the first skippers to do such a thing, Jacobson caught a lot of static. One couldn't call it flak so much as an attitude, a general disbelief. Everyone thought a woman wasn't physically capable of deck work. It was a male culture, and a man's work. The attitude shared by many of the fishermen Susey encountered was a direct "throwback to the Middle Ages," she recalls. But Susey remained.

During her first trip out as a half-share crewwoman, she worked twenty-two hours with hardly a break. Then the weather grew dangerously rough. The size of the waves dumbfounded her. Susey was terrified.

"You've been doing this for seven years?" she asked her boyfriend, Danny O'Malley. He nodded. "Hey, I'm telling you, as soon as this damned ship gets back to port, I'm getting off. And I'm not coming back out here again, I can tell you that!"

Susey was sure of it: These men were out of their ever-loving, oceangoing minds! The storm passed, though, and when three days later the *Atlantico* returned to port and Susey collected two thousand dollars for her efforts, she thought again. The *Atlantico* would be leaving the following morning at first light. The job was still hers.

Then, as she held the check, she made an astounding discovery. She noticed that not one cent had been deducted from her check—not IRS, nor Social Security, nor Employment Security, nor union fees, nor state taxes. The government looked upon fishermen (and fisherwomen) as self-employed workers.

Susey was galvanized. At long last, she sensed financial independence. She was young and single, and dependent upon no one. There was no describing such a feeling. When the ship left again at dawn, Susey was on it.

On deck at sea, Susey's "secret weapon" was a stubborn refusal to let the physical demands of the work defeat her. Through the weeks and months of work it took to get where she had, she had come to believe that it was "mostly mental."

"Move quick, and keep moving!" was her motto on deck. Whenever her back knotted unbearably or her knees grew shaky with fatigue and threatened to collapse from under her, she would recall what she had been through and what she had left. She would remember the long hot months in the Sacramento Valley, and she could picture the white stucco greasy spoon café where she had waited tables for twelve sweaty hours a day, enduring rudeness and ten-cent tips with the same philosophic smile.

It was during her second trip to sea aboard the *Atlantico* that Susey was forced to prove herself. They were stacking gear (hauling crab pots on board and storing them on deck) in heavy seas off the Portlock Banks when by a quirk of chance and circumstance, she found herself alone on the pitching, rolling deck, clinging to one of the giant eight-foot-tall, 750-pound crab pots. With the rest of the crew involved in coiling and other important matters, Susey knew she was faced with a decision.

She either could cling to the giant crab pot and wait for help or she could try pushing it across the deck by herself. At the time, she recalls, the pitch and roll of the boat was "perfect for pushing" if, that is, one knew when to push and how to go about it. With her adrenaline pumping, Susey made her decision. Bending forward at the waist, she stooped under the horizontal middle support bar of the pot and moved forward until it came to rest behind her head and across her shoulders. Like a weight lifter, she squatted low and, most important, timed her next move perfectly with the roll of the boat. At the critical moment, she pushed with everything she had.

With her legs pumping hard, Susey screamed as she drove ahead. She was sure someone had joined her in the effort and was now behind her, pushing with her. But when she stopped, she found herself standing at the far end of the deck. And when she stood up and glanced around her, she found she was all alone. She had pushed the largest of Alaskan king crab pots the entire length of the deck, and she had done it by herself.

It was a feat that did not go unnoticed, for when she raised up and turned to look around, she saw her male crewmates standing silent and frozen in place, staring dumbfoundedly at her from the opposite end of the deck.

Toward the end of the trip, one crewman came up to her and remarked, "You know, we didn't think you would be able to hold your own like you did. You really did good, young lady. Honestly, we didn't think you'd make it!"

As the *Atlantico* drew near town, deck boss Danny O'Malley yelled to her. "You know, Susey," he said, "they ruined a hell of a man when they cut the balls off you!" O'Malley never was one to tiptoe around a thing.

By the time the *Atlantico* job ended and the injured crewman had returned,

Susey had built a reputation for herself throughout the fleet. She was coordinated and strong, knew the basic knots, and was easy to be around.

"She did a heck of a job for us!" skipper Bill Jacobson recalls. "She always had the ability to see work that needed to be done. And she was dependable, too."

Susey has never forgotten her next deckhand job, a full-share position on a tanner crab boat in the waters near Chignik, on the Alaska Peninsula.

From the outset, the skipper had been short-tempered, the fishing poor, and the January weather as bad as it gets. Caught far from shore by thirty-knot winds packing a freezer-box temperature of minus five degrees Fahrenheit, the blowing spray built up steadily on the ship's cabin and rigging. Finally, they headed for shelter.

Susey broke thick ice from their anchor winch and dropped anchor on the semiprotected lee side of the Semidi (pronounced sem-ee-dee) Islands, some 110 miles from shore. Susey had never seen such an inhospitable and lonely place as this tiny windblown stack of rocks surrounded by tens of thousands of square miles of rugged Gulf of Alaska sea. The area offered their guests a poor and rocky bottom for lodging an anchor, limited protection from storms, and a rookery of sea lions for company.

Less than a day later, when the skipper ordered that the anchor be hoisted, Susey and her crewmates obeyed. But they quickly discovered that the icing conditions away from the partial shelter of the islands had only grown worse. It was a time to hole up, the deckhands agreed—a time to run for cover and wait out the storm.

As expected, ice began forming on the ship's deck and superstructure. Gradually, the vessel began to rock sullenly from side to side, and its recovery became slow and uncertain. "We figured our jobs weren't worth anybody's lives, or broken bones," Susey recalls. "And when we tried to work, we nearly got a man killed. Besides, the skipper hadn't been in contact with anybody on the VHF, and we didn't even have a single sideband radio on board!"

When, after the better part of a day, the storm winds and building ice showed no signs of letting up, Susey and her crewmates went inside and confronted the skipper.

"Look," said Susey. "This is crazy! We're going to get someone hurt out there and we're not catching anything. We need time to break off the ice we're carrying. Let's anchor up and wait for the weather to come down." Her crewmates nodded in agreement.

When the skipper refused, Susey and her crewmates pulled off a mutiny of sorts. They refused to work. At this, the skipper came unglued. "You sons of bitches are the worst bunch of crewmen I've ever had aboard this ship!" he shouted. When they still refused to budge and, instead, asked to be taken to the nearest port, he retorted: "You bastards just lost your bonus! You can forget it!"

The skipper went below then and shut down the main engine. He was going to lie down, he said. The boat could just drift. They could wake him when they were ready to go to work.

"And this, with the boat icing heavily all the while!" recalls Susey.

But the crew held fast, and finally the skipper realized that he had no alternative and returned them to a floating catcher-processor in Chignik Bay, where Susey and her crewmates quit on the spot.

Then came the job aboard the *Royal Quarry*. In her first ten days working for Mike Jones during king crab season, she earned seven thousand dollars. It was an incredible experience for a twenty-year-old single woman.

Along with Susey Wagner, myself, and a quiet young man from Seattle, curly-headed Steve Calhoun rounded out the *Royal Quarry* deck crew. Originally from Yuba City, California, Calhoun had been a standout high school football player who'd struck up a friendship with our skipper while attending college at Oregon State University in Corvallis.

Like thousands of youngsters who had come before him, Calhoun arrived in Kodiak without a job. Though it was officially spring, it was winter-cold outside. With only forty dollars to his name, he bought a sweater that read "KODIAK ISN'T THE END OF THE WORLD, BUT YOU CAN SEE IT FROM HERE!" That night, he spent twenty-four dollars of his money on a hotel room. Outside, it was snowing.

The next day, he found a job in a cannery on Cannery Row. He was given a couch in a boardinghouse filled with Filipino men and proceeded to work day and night. Some weeks he put in more than one hundred hours! At $2.90 an hour, he knew he wouldn't get rich, but it was worth his time. It would have to do until something better came his way.

Still, Calhoun never forgot why he had come. He had an itch to go to sea, to become a fisherman, he recalls. It was a desire that never left him. After working sixteen hours a day, often seven days a week, he found he had little time, and even less energy, to beat the docks and look for a crewman's berth on board one of the shrimp boats. Also, he soon discovered that fishermen didn't hang out with cannery workers. Their paths rarely seemed to cross.

A few weeks later, the local Kodiak fleet of shrimpers went on strike. So, in his off-hours, Calhoun started hanging out in the local bars and sharing a few brews with the striking crewmen. Unlike Calhoun's mundane existence, their lives seemed exciting. They worked hard, made incredible amounts of money, and told adventurous and occasionally harrowing tales. He felt at home in their company, and drawn to their wild and independent way of life.

Then one day, he learned that one of his drinking buddies had jumped ship off the *Debbie D* and had hired on with another shrimp boat.

"The moment I heard the job was open," he recalled, "I ran down and presented myself to the skipper."

"You have any experience?" the man wanted to know.

"A little," lied Calhoun.

"Well, do you know how to take a wheel watch?"

Calhoun's mind raced. He had no idea what the skipper was talking about. "Sure!" he bluffed. "I can do a little of that."

There was a dubious look on the skipper's face. He hesitated. "Well," he said finally, "you might as well throw your gear aboard."

Calhoun hardly could believe his ears. In the season ahead, his boat caught and delivered 2 million pounds of shrimp. Calhoun's share alone would top $25,000.

Affectionately called "Jonesy," our skipper was only twenty-three when he first came to Alaska. He'd worked summers hustling along the docks and processing lines of Newport, Oregon, ever since he was just a youngster, he recalls. He saved his money, and built a reputation as a worker and as a young man upon whom a skipper could depend.

He came to Alaska aboard the *Bernedet,* a Newport, Oregon–based shrimp boat, and on the first day of fishing just outside of Chiniak Bay, the *Bernedet* netted forty thousand pounds of shrimp in a single tow. "We plugged [filled] the boat in about a day," he recalls.

Each summer, then, Mike Jones fished for shrimp out of Kodiak; and each September, he returned to Oregon State University to study and wrestle. This tireless and highly charged future skipper and owner of the *Royal Quarry* would go on to win the Pacific 8 wrestling title two years running.

After graduating from OSU, Jones worked out of Kodiak as a biologist for the Alaska Department of Fish and Game. But eventually he returned to the sea, working for Teddy Painter, an exceptional fisherman (also from Newport) who owned and operated the *Buccaneer* out of Kodiak.

Jones worked for two straight years putting away money. He saved every dime, and he kept his eye on the market, learning all he could about crabs and the boats that fished them. "I ended up with about thirty-five thousand dollars, and then I ran into Lyle Negas at the B & B Cannery there in Kodiak. He owned the *Royal Quarry.*"

Jones knew that there would be endless repairs to make and gear to buy. "But, you know, I didn't have much to lose," he adds. "Most of the guys I knew . . . were spending it buying this or that . . . things . . . doping or drinking and partying. I liked to have fun but I knew it wasn't for me. I saw a good opportunity . . . a chance . . . and I decided to take it."

THREE

Back aboard the *Royal Quarry*, I was hiding in the refuge of my bunk when the vessel sounds echoing through the boat's interior grew suddenly muffled. When I looked up from my horizontal misery, I saw a short, sinewy figure standing in the doorway. It was Jonesy, and his words were brief. "We're on the gear!" he said, disappearing up the stairs into the wheelhouse.

Ashamed at my unrelenting seasickness, I rose without looking and clipped my head on Calhoun's overhead bunk. I staggered sideways, my heart pounding in wild surges as I gasped for air and reached for my work clothes.

Outside, I plunged ahead on uncertain legs across a rudely gyrating deck and quickly discovered that maintaining one's balance on the shifting, heaving surface was a labor all its own. We had arrived at our pot gear at just after dawn, and a more cold and bleak and gray wasteland of sea and sky I had never seen.

Our first chore was basic and critical: to piece together our entire back deck. "It's no big thing," Susey informed me. "We have to do it every time the weather kicks up. The waves break across the back deck here and wash the pieces around." I followed Susey's pointing arm and spied more than a dozen odd-sized chunks of the deck washed into a broken pile in the more or less middle of the deck.

We used crowbars to separate the pieces and, bending to the task, physically pulled them back in place and slid them together. The reconstruction went much faster than I had imagined; less than fifteen minutes later, the primitive deck lay restored beneath our feet and we were ready to fish.

We had come here to catch tanner crab, which the Japanese were buying up as fast as we could catch them. What didn't get shipped to the Orient found its way to the restaurants and dinner tables of other Americans living more reasonable lives in the South 48. These crab were commonly marketed in restaurants and steak houses as snow crab.

It has been only in recent years that canneries would even buy tanner crab. Previously, there had been no economic way to extract the succulent meat from the long, spindly legs. And at eight or ten cents a pound, few fishermen found tanner crab fishing worth their time. But now the processing plants along Kodiak's Cannery Row were offering up to forty cents a pound.

Longtime Kodiak Island fishermen could tell you that for more than a decade their king crab pots had risen to the surface crammed tight with tanner crab. The wandering tan-colored critters were everywhere, making king crab fishing in certain areas impossible. And the king crab fishermen cursed them with every obscenity known to man.

But tanner crab had grown steadily more valuable and noticeably more elusive. Big crab ships with experienced Alaskan skippers were returning to port with only partial loads. Even the best of fishermen were finding that the crab were scattered and much more temperamental than they could ever recall. The tanner crab were going on and off the "bite" with surprising moodiness, so it was with high hope but guarded expectations that we prepared to pull the first of the *Royal Quarry*'s crab pots.

Like most greenhorns entering the trade, I had only a superficial understanding of the process of crab fishing. I soon learned, however, that, as with most productive systems involving the manipulation of weight, leverage is the key.

The process involves a six-by-six-by-three-foot crab pot made of steel and webbing and weighing 450 pounds empty. Ours, commonly referred to by fishermen as "six-bys," were the smallest and lightest of the pots being used by the fleet. The more commonly used crab pots ranged in size from seven (nicknamed "seven-bys") to eight feet ("eight-bys") and weighed seven hundred to eight hundred pounds empty. Even our pots weighed nearly a quarter of a ton empty, and when loaded with tanner crab, their gross weight as they swung aboard could exceed a thousand pounds—enough, in heavy seas, to drag a man across the deck or crush him to death if mishandled.

These "six-bys" are baited with chopped-up bits of frozen herring and launched over the side by a flat tablelike steel "rack." This is accomplished with the use of the hydraulically powered rack. When it tilts up sharply, the crab pot slides over the side, much the way high-seas burials are performed.

The pots come to rest on the ocean floor, usually three hundred to six hundred feet below, and are connected to the surface by a ⅝-inch line secured to a set of three pot buoys. The arrangement, color, and markings of those buoys differ with each vessel, but they generally include one Styrofoam buoy and two air-filled plastic buoys that are fluorescent orange in color. The Styrofoam buoy is called a "sea lion" buoy because it will remain afloat even if sea lions, who often delight in such antics, bite and pop the two air-filled buoys.

To harvest the crab, we first identify the pot as ours. The last buoy in our

23

gear setup is marked with the big black hand-stenciled letters *R.Q.* on either side of it. A hook line is used to snag the floating buoy setup and pull them aboard. Then the line is placed in the pinch of the powerful hydraulic steel block and winched to the surface. The pot is then craned aboard, guided into position in the "launcher," or "rack."

If the fishing is good and the pot carries a profitable load of crab, it is quickly rebaited and dumped back overboard. If the fishing is poor, the pot is stacked on deck in tightly bound rows for transport to other fishing grounds. Using this system, an experienced crew working in good weather can process between fourteen and eighteen pots an hour. In bad weather, the pace falls to ten or fewer pots per hour, depending upon the severity of the storm.

As our ship maneuvered to the first trio or set of buoys, I could see them tossing and bouncing in the rough gray seas. Each time a wave rolled in over them, they disappeared beneath the surface.

Calhoun had a reputation for accurate throwing, and I watched intently as he timed his movements with the downward surge of our boat, swung the hook and line like a lasso, and flung the steel-pronged grappling hook and trailing line out over the buoys. The grappling hook cleared by no more than a yard the crab line connecting the buoys. Calhoun paused momentarily to allow the hook to sink, then jerked back on the trailing line like a fisherman setting his hook. In one continuous motion, he snagged the submerged buoy line, pulled the setup back to the boat, and flipped the buoys on deck.

As I pulled the buoys out of the way, Susey grabbed a section of the crab-pot line and flipped it into the stainless-steel crevasse of the hydraulic block, signaling Calhoun with a quick nod of her head. Then Calhoun cranked on a lever and the motionless block dangling in front of Susey leapt to life. Undaunted by the sudden flood of line spewing aboard, Susey flew into action. With feet braced wide, she made several indecipherable flips of her right hand and, as if by magic, managed to clear the deck in front of her. She then guided the line aboard in a casual hand-over-hand fashion, coiling it as it came into a perfectly shaped oval bundle at her feet.

We were fishing in 86 fathoms (516 feet) at the time, and in no more than two minutes, Susey's tight coil of line grew to a height of nearly three feet.

When the pot's bridle broke the surface, Calhoun flipped a lever, cutting power to the hydraulic block, which froze the pot in midair alongside the boat. In quick cadence then, Susey grabbed the two-foot-long steel J-hook dangling from a long steel boom overhead and snagged the bridle as close to the center as possible. As Calhoun winched the pot up over the side, I heard a collective groan rise from my crewmates.

"Damn," piped Calhoun as we swung it into place in the ship's pot rack, "giving them a two-day soak the way we did, I thought they'd fish better than this." He seemed deeply disappointed.

"We can do better than this elsewhere," said Jonesy over the deck speaker. His voice was full of disgust. "Let's pick up the rest of the pots from here on, or at least until things get a whole lot better."

24

Calhoun returned to his position beside the steel "tree" then and, working the control levers, governed the speed of our whining block. Built of shiny stainless steel and hung from the end of a curved chunk of five-inch pipe mounted inside a pivoting stand, our block weighed several hundred pounds and could walk through a three-thousand-pound test line with a casual indifference.

As the pot landed in the launcher, Calhoun leapt to the pot, pointed to my side, and yelled, "Door ties!"

I raced to untie the thin yellow line holding my side of the pot's door shut, finished a distant second to Calhoun, and joined him in flipping the door open. We then raced to remove the tanner crab and old bait. If the fishing had been good, I was told, Susey would have tossed a twenty-five-fathom "shot" of line on top of the pot, now sitting at a forty-five-degree angle in the pot rack. Then the pot would have been launched back overboard with the remaining shots of line and the buoy setup in close pursuit.

But for now, the fishing was poor, and as I watched, Susey bent and began tying the coiled pile of crab-pot line together. While Calhoun and I held the door open, she stooped and lifted the weighty yard-deep bundle to her waist. Then, in one motion, she pivoted violently at the hips, grunted aloud, and managed to toss all six hundred feet of line inside the crab pot in a single effort. One, two, three, the buoy setup followed. Then Calhoun flipped the door closed and we tied it securely in place. On the next roll of the boat, Calhoun and I hoisted the ponderous five-hundred-pound pot upright.

"Now, here's something you just might be able to do!" Calhoun called to me as we held on to the pot to keep it upright in the bouncing seas. "I mean, you can *push,* can't you?" he ribbed. "You squat under that crossbar there, and Susey and I'll help guide you. We'll shove it right over in that far corner."

Seasick and feeling weak as an infant, I lowered my six-foot, 245-pound frame so that my shoulders rested against the middle support bar. With my head inadvertently stuck inside the crab pot's tunnel, I dug down and drove ahead.

The effort was an act of faith, for I pushed not out of the physical death that for the moment owned me but out of the memory of the strength I once had possessed. With a mixture of surprise and even joy, I felt the pot immediately begin to move, and the slow, dogged shove for which I had prepared myself quickly accelerated into an across-the-deck sprint. Seconds later, the pot slammed into the far corner.

"Okay! All right!" shouted Calhoun, rushing to tie the pot to the railing. Susey was laughing excitedly.

"All's I want to know now is how much is it going to cost me!" she chided. "I mean, how much do you *eat* when you push like that?" She patted me on the back like a coach giving her approval.

Calhoun took the time to shake my hand, then raced across the deck to give the throwing hook a toss. The next set of buoys, Jonesy had warned us, was already upon us.

It was a splendid feeling, the sudden esteem of being able to help my fellow crew members. Over the next several hours, I felt my seasickness diminish and,

25

as I leaned to and heaved ho, I felt a renewed confidence—for each new crab pot we hauled aboard seemed to slide across the deck ahead of me as easily as a sled on ice.

For the next twenty pots, the fishing remained poor and we averaged fewer than thirty tanner crab per pot. As Susey and Calhoun worked together to bring each pot to the surface, I busied myself chopping bait, filling bait jars, and measuring the backs of crab with a hand-held stick to make sure they were of legal width, all the while praying silently for our luck to change.

To our delight, that afternoon the fishing picked up sharply. When crab pots began arriving at the surface packed with a knee-deep pile of more than one hundred tanner crabs, we hastened to dump overboard the twenty or so crab pots we had spent the morning picking up.

For the next twenty minutes, we used the hydraulic boom hook and cable to drag the pots, one by one, from the stack of pots on the stern to the pot rack. Then Calhoun and I would lay the pot over onto the rack and prepare it for launching. With Jonesy at the helm, the *Royal Quarry* rose, staggered, and fell through rough and building seas as she fought to maintain her course. Regardless, as she powered up and over the steep waves, we continued to dump our crab pots overboard at the respectable pace of one pot per minute.

Then, returning to the string of hot fishing gear, we greeted each emerging crab pot with reborn enthusiasm. In an ever-more-coordinated effort, we rushed to remove the crab, rebait the pot, and dump it back over the side in the shortest amount of time possible. Then the female crab had to be sorted and tossed back overboard, and the males counted. As we listened intently each time Calhoun relayed the catch count for each pot over the two-way deck speaker to our skipper, our spirits soared.

On deck, crewmen became mesmerized by the rhythm of work and the sounds of the life. One took in the incessant cries of hundreds of sea gulls battling over discarded scraps of herring bait, the unexpected "pock" and jolt of a rogue wave catching us amidships, the belching growl of our diesel engine, the cold snap of unexpected spray on the face, the pop and crackle of crab line feeding through the block, and the metallic chink of tanner crab bodies clattering against the steel chute leading down into our live tank below deck.

I spied my first king crab that afternoon. It had forced its way inside, squeezing past the tanner boards—planks wired across the two entrances to the crab pots to allow the passage of the relatively thin tanner crab, but usually not the husky king crab. It was sprawled menacingly across the bottom of the pot. It was red and spiny and looked about four feet wide. Its right pincher appeared to be about the size of a man's fist. When I hesitated to pick it up, Susey said, "Well, either grab him and we'll cook him, or toss him back. He won't be worth a thing to us in town until next fall."

"But he sures does look healthy, now doesn't he?" put in Calhoun. Bending down into the pot, he grabbed the creature like a man gripping a steering wheel, and, with one chunky crab leg in each hand, he hoisted the bulky twelve-pound crustacean from the pot and dropped it back over the side.

We'd hoped to run through all eighty-eight of our crab pots that day, but by the time we had worked our way through pot number forty-six, the wind and mounting seas forced us off the deck. Then Jonesy spun the large wooden wheel hard over and we headed for cover.

Susey explained that it was not uncommon for crab boats and their crews to spend entire weeks holed up, hiding out from weather too tough to fish. That night, wtih high-wind warnings posted, we slipped through the rock-bordered narrow entrance into Lazy Bay, a small land-encircled inlet offering partial protection from the steady typhoonlike winds and tumultuous, pounding seas. Here we came upon a covey of perhaps twenty crab boats swinging heavily on anchor and drifting sharply back and forth across the bay. Their straining anchor chains extended forward from the point of each bow, slanting down and disappearing into the rough black bay water. Soon, the violent wind velocities of the storm moving in over us smeared the inlet's face into white streaks of foam. Eventually, silver-gray clouds of spray began to kick up, scattering across the bay in a misty vapor.

Shortly, I heard the sharp steel rattle of our anchor chain as it played overboard, and I knew that the day was at end. Exhausted from some twenty hours of seasickness and an all-ahead effort on deck, and with eighty-knot winds howling around the rigid steel form of our wheelhouse, I gave thanks to my higher power for having survived the first of my sea trials. I had no idea, I said to myself with a final sigh. Then I fell off to sleep, dead away.

FOUR

Four hours later, Susey woke me. "You're up on anchor watch, Spike," she said, shaking me awake.

I dressed and climbed into the wheelhouse, listening through a fog of fatigue as Susey tried to convey how alert and responsible I needed to be during my wheel watch. "It's your responsibility to make sure we don't drag anchor," she began. "If we do begin to slide backward in this strong wind, we'll be on the rocks back there in about two minutes. So you go below and wake Jonesy. If you've got some question about the radar or the weather, then get one of us up. The point is, if you think something is going wrong, call for help." I nodded my understanding.

"You can tell if we're drifting back or not in a couple of ways. The first is to keep an eye on the other ships nearby. The most accurate way, though, is to keep track of how far we are from the shore by constantly checking the radar. Just remember that when I go below and we're all down in our bunks asleep and you're up here alone, the boat, our lives, and the whole damned season are in your hands."

Susey went on to explain that whenever she was alone on watch and tired beyond anything most working people ever know, she tried to recall the seriousness of her duty and to think of the men who were equally as tired. It was a way of charging herself up. Sometimes, when she felt especially good, she doubled up on her watch (taking two shifts) so that the others could sleep on. Other times, when she felt sleep stalking her, she'd remind herself that at that moment everyone on board was depending upon and trusting in her. Then she would

28

jump up and pace the floor and periodically lean out the wheelhouse window to let the cold wind chill her awake.

For two hours, fierce winds buffeted us and my eyes leapt between the fluorescent green lines of our radar screen and the coal black void of night all around.

For the next two days, we remained stormbound in Lazy Bay. Then, on the morning of the third day, Jonesy shook us awake. "The weather's come down a little," he assured us. "We're gonna go take a look."

Assaulted by the fierce winds of yet another barren-looking morning, and while the rest of the fleet rested, Calhoun and I climbed out onto the bow. "I can't see where the wind has come down at all," I yelled into the wind, shaking violently beneath the thick wool folds of my Navy pea jacket and black stocking cap. Without further discussion, however, Calhoun activated the hydraulic winch and began bringing the anchor aboard. Then as we made our way back around the wheelhouse, Jonesy swung the *Royal Quarry* into the wind, pivoted her around, and began idling out of the harbor. And while the crews of more than a score of sleeping crab boats swung silently on anchor, we weaved our way through the slumbering fleet. Finally, we slipped past the reef and rocks at the mouth of Lazy Bay and, swinging to starboard, moved out into the wind-raked waters of Alitak Bay.

It was a four-hour run out to our gear, and though our sleepy-eyed crew entertained grave doubts as to our skipper's judgment in such weather, we prepared to do battle in heavy seas.

Strangely, several hours running time from the mouth of Lazy Bay, the waters began to calm and by the time we came upon our buoys dancing on the surface more than twenty miles offshore, the clouds had parted and faint white shafts of winter light filtered down on a mercury-colored sea. It was still a little rough and unpleasant, but we would be able to work.

"How in the hell did he know?" I asked, putting it to no one in particular, as, once again, Calhoun, Susey, and I pieced together the scattered remnants of our back deck.

"This isn't the first time I've seen Jonesy call it like this," said Susey.

From the outset, that morning, our crab pots rose heavy with legal-sized, web-stretching piles of tanner crab inside. They tumbled from our pots and chinked against the side of our metal live-tank chute like silver dollars from a one-armed bandit. In only the first ten pots, we caught more than a ton of crab.

Breakfast and lunchtime came and went without notice, nor were there any complaints. For while it was still a cold and dangerous job, pulling gear packed with such beautiful writhing throngs of crab had its own reward. I felt expectant, like a gambler watching the final roll of his dice. As the crab pots broke into view on the surface, they produced a roller coaster of emotions. Each plugged pot sent us into an exuberant dance, while, only minutes later, a blank pot would pitch us into black despair.

Contrary to the hypothermic effect of the wind and spray, my eyes burned as rivulets of sweat and saltwater raced down my forehead and spilled into them.

So I tried the trick I'd learned in the logging camps of southeast Alaska: I applied a single swipe of Vaseline across the arch of each eyebrow to rechannel the salty rivers.

All day as we worked, tireless flocks of sea gulls dove and squabbled over the smelly jars of used herring bait emptied overboard. Occasionally, when we'd swing a crab pot aboard, we'd be greeted by the ugly snarling stare of a wolf eel, or the skeletal remains of a gray cod, black cod, or halibut that had unwittingly found its way into the strange dead-end maze of our submerged trap in the darkness on the ocean bottom below. Unable to find their way out, they often fell prey to what are commonly referred to as sand fleas. These maggot-sized shrimp-like creatures swarm to the trapped, dead, or dying fish and strip them of their flesh like millions of tiny piranha. They pick the carcasses so clean that the bones look as though they had been boiled.

I started at the sound of an explosion and twisted about in time to see the hydraulic block dancing from the end of the boom. A limp and weightless crab-pot line peeled off it and spilled helter-skelter onto the deck. Calhoun cursed vehemently. The line had parted under the load as the ship surged over a wave.

The line had either been worn and passed undetected, or chafed by the line from another vessel's crab pot placed too close to ours, or the pot had been winched up off the ocean floor in too great a hurry. Regardless, the crab pot and, more importantly, all the crab it would have caught in the future were lost forever.

While working around numerous logging camps in the South 48 and southeast Alaska, I'd grown accustomed to what are widely known as "rigging fits." In reality, they are nothing more than adult temper tantrums, but are rated by old-timers on any number of varying scales, from the diverseness of cuss words used to curse the person, place, or predicament, to the wild impromptu dances that often accompanied them. And so I waited for a violent reaction to come from the wheelhouse. But none came. Jonesy's response was dead silence.

The responsibility to check the gear—which included all the knots, shots of line, and buoys that came aboard—was ours, and as deckhands we could only squirm in our skipper's silence. There was nothing to react to, no way to explain, rationalize, or argue. And, in the end, his silence was more effective than a beating.

Throughout the wet and windy day, as each new crab pot was wrestled aboard, Susey routinely separated the shots of line, checking the knots and line for chafing. Now and then, she was forced to break the frayed and ragged-looking knots with a hammer. If she called out "I need a new shot here," I'd hurry over to the pile of line stacked against the wheelhouse and toss a twenty-five-fathom shot of line to her. She would then tie new shots of line together, toss me the old shot, and prepare the crab pot for launching.

After the buoys of one crab pot had been brought safely on board, Susey invited me to spell her at coiling. With cocksure confidence, I positioned myself just as I had seen Susey do. With my legs braced wide, I threw the line in the block and gave Calhoun an authoritative nod to begin reeling in the crab pot.

I gripped the line tightly, pulled mightily on it, and tried to wrestle it aboard. The first few loops of the coil had just begun to form at my feet when the thick yellow polyester crab line leapt from my hands and began spewing from the block and between my feet and around my legs in a wild and uncontrollable jumble. Calhoun stopped the block and I backed away in defeat.

Without hesitation, Susey moved into my abandoned position, plunged both hands into the center of the tangled pile of line, and with several flashes of her right arm managed to clear the line and bring an oval order to what I had deemed a hopeless mess. Then, flipping the line like a lasso, she proceeded to coil it as fast as our Yaquina Bay block could spin it aboard. And as she coiled, each new loop landed in a perfect oval form at her feet.

With buoyant optimism and an almost saintly patience, Susey turned back to me as she coiled and said, "See? There's nothing to it, Spike. But it does take a little time to pick it up. So just watch me for a while."

I watched with newfound admiration as the line flew through her hands. As quickly as each new loop formed in her hands, she flipped it down onto the growing pile of crab-pot line at her feet, tossing it to within a fraction of an inch of the one that had preceded it. Susey's hands flashed here and there as she worked, and as they moved, her eyes took in the condition of the coil, the direction and strain on the line, and the angle of the spray flying off the spinning block. She did it quickly, with a kind of casual indifference, as if by instinct.

As I stood there, it dawned on me that I was being watched, tutored, cussed out, fussed over, and ordered about by a 135-pound former short-order waitress. What I was experiencing went against the masculine Alaskan myth that I secretly believed in. And I felt something ancient and unreasonable welling up in me. Oh, God, how I hated liking that young woman!

FIVE

Over the next few days, the February wind and weather improved, but the best the sun could do was to illuminate the formless cloud cover and drifting fog in a cold and silvery haze. Through repeated attempts and failures at coiling, and much personalized coaching by Susey, my coiling improved to where the shots I coiled on board were acceptable if not yet admirable.

Three times during the week, we returned to Lazy Bay for rest and a hot cooked meal, and allowed our baited pots time to lure more crab. Throughout our last turn through our pots, we baited them heavily with whole codfish and jars of chopped-up herring and returned them to the sea. Then we turned toward home.

Four complete passes through our gear that week plugged our single tank and netted us roughly 18,500 tanner crab, for a total catch of more than 48,000 pounds.

With a new storm reportedly driving into the east side of Kodiak Island along the entire length of the route we'd followed out to the fishing grounds, Jonesy decided to return to town by traveling up the opposite side of the island. But by the time we rounded Cape Ikolik, the wind had come around and we found ourselves caught in the pitch-darkness of a cold and wind-raked February night. With 110 miles of steep and wintery waves standing between us and Whale Pass and the village of Kodiak, we plunged ahead, pounding our way up and over and through unrelenting mountains of wind-driven waves. We pressed ahead at full speed but could make no better than seven knots as we fought our way along the entire length of Kodiak Island.

It was daylight when we rounded Cape Uganik and turned into Kupreanof Strait. There, between the steep mountain ridges bordering the strait on either side, the seas calmed dramatically. The nightlong commotion was over, and the *Royal Quarry* motored ahead unchallenged toward Kodiak.

That evening, as we drew closer to town, a fresh new spirit surfaced in our crew. Clad in a clean change of clothes, Calhoun and Jonesy joked freely and without worry. Then I came upon Susey blow-drying her hair in front of the galley mirror. Though still in boots, she wore lipstick and a stylish pair of snug pants that did little to conceal a substantial figure.

I couldn't get this young fireball out of my mind. The night before, she had walked past me wearing only her panties. Eight days at sea, I thought, and she wears only her panties. I could see crisp black pubic hair pressing naturally against the sheer semitransparent fabric of her pink undergarment, and a full and powerful bottom moving rhythmically as she passed. "Susey, will you put on some pants? Heck, I'd be glad to help you," I offered quietly, hinting at a private pact. She stopped, smiled, and patted me on my cheek. "Poor baby," she said as she stepped into her stateroom and closed the door behind her.

Despite the friendly rebuff, I felt antsy with anticipation. That night, I'd throw back a dozen or more straight shots of raw eye-watering tequila, my own slow form of hemlock. Standing on the back deck of the *Royal Quarry*, breathing in the damp, cool, sea air, I could already feel the raspy burn of it swilling down and the lingering cool of evaporating alcohol flooding through my sinuses. In the glitter of the waterfront lights, we tied up at the B & B Cannery docks. I was helping Calhoun sling the stern hawser line around a piling when Hungarian Joe, the cannery foreman, appeared on the edge of the dock high above us.

"So!" he yelled down at us in his strong accent. "You go and do good, yes? No? I hear you do good!"

Calhoun waved. "We're nearly plugged again, Joe! I'm not going to complain. We got our asses kicked during the first part of the trip, but I'm not going to complain."

Jonesy walked out on the back deck then. "We'll pop the cover anytime you want them!" he yelled, looking up at Joe.

"In zee morning!" Joe called back. "Vee got no more boats coming tonight. And zee crews . . . they are chasing zee women. Vee take your crabs first ting in zee morning!"

With his hands braced on the kidney area of each side of his back, he stood looking down at me. "Vell, how did you like it zee trip?" he asked, all the while sporting a good-natured grin a yard wide. "You tink you want to be big fisherman? Yes? No? Jonesy, ahhh, he know how to catch zee crab. I know this now, goddamn it!"

I smiled up at Joe, nodded, shouldered my packboard, and began scaling the dock ladder. Secretly, I hoped that I could hold out financially long enough to locate a crewman's berth on a Kodiak crab boat. I still lacked a job. For, just as Jonesy had explained before the trip began, I had no crew share coming, nor

33

had anyone from the crew of the *Royal Quarry* quit or been fired. I had agreed to make the trip for free, and therefore I did not feel cheated.

With three crewmen still employed on the back deck of his vessel, Jonesy made it clear that he had no berth to offer me. Still, it was no secret that there had been a noticeable tension among the deckhand from Seattle and Susey and Calhoun, and several times during the trip they had been openly vocal about it.

I had never been one to wish bad times on another. I don't believe I've ever taken pleasure in another man's pain or danced on another's grave. But it has been my experience in brutally physical roughneck trades such as logging, oil-field work, and now fishing that the man who is strong and knows how to work will eventually find a place.

Besides, I told myself, with Jonesy's help I'd come by an irreplaceable block of experience. I had been exposed to the working life of a deckhand aboard a crab boat out on the high seas and felt that I now knew enough to contribute meaningfully as a crewman at sea. I had learned it was a wild and unpredictable existence, and that no two days were alike. I'd learned that a man who wanted it badly enough could work through even the worst bouts of seasickness if he reached down far enough. And now, more than ever, I wanted to become a part of this life.

That night, Jonesy treated me to a large steak with all the trimmings at Solly's. The restaurant was packed with hungry fishermen and filled with the rustle and talk of fishing. Noise flooded in constantly from the nearby bar as we talked of the trip and our catch. After we had eaten, Jonesy extended his fist, palm down, across the table and dropped a small scroll of cash into my hand.

"Here, Spike," he said. "I really appreciate your going out with us this last trip. You really helped us out, and it was your first time to sea. You've got to feel good about that."

When I apologized for being seasick so much of the time, he smiled. "Everybody gets seasick. Everybody. And if you do get sick, well, that's your problem. It becomes my problem when you let it affect your work, but you didn't let that happen."

He stood up to leave and turned back to me. "I'd be patient and hang around if I were you. Don't go anywhere. I may need you real soon." He paused. "I'm going to take the man from Seattle out with us on one more trip and we'll just see how things work out. Susey and Calhoun are getting awfully testy. But, God, I hate to have to let any man go."

After he'd departed, I opened my hand and unrolled the small, tight bundle of money. It was composed of two crisp one-hundred-dollar bills.

Feeling instantly affluent, I bolted down the rest of my eighteen-ounce T-bone steak and hurried off to pay back the twenty dollars I owed Calhoun. I rode to his place in the warm extravagance of a taxi. And as the cold winter scenes of wind and ice fled past, I could never remember feeling better.

When the *Royal Quarry* left on its next journey, I remained in port, taking loan of Steve Calhoun's apartment while he was away. With the full brunt of an

Alaskan winter gripping the countryside and the *Royal Quarry*'s return a week or more off, I felt thankful for the warmth and protection of its walls, the availability of food in the cupboards, and the security of a few dollars in hand.

Five days later, I was still patiently awaiting the *Royal Quarry*'s return when I received a call from Carl Bock, skipper and owner of the crab boat *Van Elliot*. Our paths had crossed on the waterfront and in Lazy Bay the week before. He knew I was looking for work and had called because he needed a crewman to fill in for about a week for one of his crewmen.

"You'll probably make between one thousand dollars and fifteen hundred dollars," he told me. "Do you want to go?"

My mind raced. I was broke, in debt, and half a continent from home. A thousand or more dollars seemed a fortune, but if I took the fill-in work and went out on another boat and Jonesy returned and needed a man, would he hold the job for me? Jonesy *did* tell me not to go anywhere. Those were his own words.

"Could I call you back in a little while, Carl? I need to make a call and then I'll ring you right back."

"No problem," he replied. But he said he'd have to know, and soon.

Hanging up, I immediately called another fisherman, a young Oregonian named Jim Berg. He'd been a friend of Jonesy's and mine down at Oregon State University in Corvallis.

"Jim," I began, panting with excitement, "I just got a call from the skipper of the *Van Elliot*. He wants me to go out on a trip for him. He's running a man short. It's a fill-in trip."

"Well, go then!" shot back Berg.

"I know, I'd love to," I replied. "I sure as hell can use the money. But the *Royal Quarry* is due back in town any day now and Jonesy told me he might have a job opening up for me. I need the money bad, but if the *Quarry* returns while I'm gone and they *do* need a man, what will happen? Will Jonesy hold it open for me? Is that how it's done up here?"

"Nope," came Berg's confident reply. "He'll take the best man who's available at the time. The *Royal Quarry*'s a great little boat. It's first come, first serve in this business. If I were you, I wouldn't go."

Berg seemed sure of himself and I had nothing to lose, really. I thanked him and called Carl back and declined his offer, but I'd hardly hung up the phone when my mind seized on the decision. That night, huddling in my sleeping bag on the living room floor, I doubted my sanity. How could I have let fifteen hundred dollars slip through my hands like that?

Late the following night, a ragged-looking Steve Calhoun came trudging into his apartment, trailing snow behind him. "We just got into port," he said. Ignoring my questioning face, he walked casually to the refrigerator and took out a can of beer. Then he snapped it open, tipped it back, and took a long draw.

I couldn't stand it anymore, and when he saw my bearish approach, he started to laugh. "Don't you even want to know how we did?" he asked. He paused to belch. "Oh . . . we had a man quit the boat this trip. Jonesy said to tell you that if you want the job, it's yours."

I let out a whoop and, clenching both fists, leapt clear off the floor.

With a few dollars in my pocket and the knowledge that I'd just been hired onto one of the more successful crab boats in the entire Kodiak fleet, I spent a night out on the town. The *Royal Quarry* once again had returned with a near-full load of tanner crab. Calhoun's share had come to more than twelve hundred dollars. "Not too bad for a week's work," he said, joining me.

We began our night of escapades downtown and ended up drinking and hooting away the hours out at the Beachcomber Bar. In the late 1950s, the bar's clever owners launched their business by dragging a large passenger ship—*The Beachcomber*—several hundred feet inland from the ocean's shore and turning it into a hotel and bar, complete with food, drink, and live music. In recent years, the business had been moved a few hundred feet away into a new stucco-style building, which sports one of the longest bars in Alaska.

The new Beachcomber had a flat roof, an elevated stage, and a sunken dance floor made of a solid plate of stainless steel. It was a clean establishment and, like the Beachcomber of old, was run in an honest, straightforward fashion. The owners kept the glasses spotless, served up an honest drink at a reasonable price, and did their best to create an atmosphere of fun. Drugs were not tolerated.

When Calhoun and I entered the Beachcomber, we found it crowded with a rough and boisterous bunch of pot maulers. Every seat along a bar that stretched for perhaps eighty feet was filled. Amid the music and the boastful conversation, simple communication required that one lean close and yell. Everywhere, one could see evidence of loose change and big bills, and the flagrant abhorrence crewmen seemed to have for saving it. Amid the crowds pressing in against the bar, serious drinkers ensured a steady supply by buying a six-pack (a half dozen shots of hard liquor) at a time. As the night wore on and the money and liquor flowed, the excitement built toward a mild hysteria, and a small riot seemed always in the making.

I'd crossed such scenes before, in the oil-field watering holes of Morgan City, Louisiana, when the oilmen and the shrimp fishermen got together; in the logging town saloons of Forks, Washington; and in the waterfront bars of Ketchikan, where, on the Fourth of July, the salmon seiners and trollers and gillnet fishermen motored into port and tied up their ships, scores of logging camps shut down and their wild and unpredictable crews caught pontoon bush planes into town, and the whole lot of them crossed paths.

But there was more money here in Kodiak. Much more money. Nowhere else on earth could young laboring men earn so much. Scores of deckhands were making from $30,000 to $80,000 a year, and some as much as $100,000; others were earning even more. And yet the more they made, the wilder they behaved and the more determined they became about spending it.

Anything could happen in the bars of Kodiak. One could sense it—the feeling of risk; the probability of the prank; the sudden outburst filled with emotion, bizarre behavior, or unexpected violence. Coupled with the close ca-maraderie of hardworking friends, the enticing oblivion of alcohol and drugs, and the late-night bar hours, the life lured many crewmen from bar to bar, night after night, whenever they were in port.

Once during the week-long wait for the *Royal Quarry* to return, I witnessed the drawing of a solid ten-foot line of cocaine in plain sight on the counter of a local bar. As I watched, the signal was given and the straw-toting crowd attacked the drug, inhaling perhaps two thousand dollars' worth of the narcotic in only seconds. In one of the smaller bars, three times that night someone rang the bell hanging over the bar, signaling drinks for the house, at more than $160 a round. I learned from the bartender there that during the king crab season a few months before, one of the more successful skippers had rung the bell fourteen times, at $174 per round. Each time the house would cheer their approval, and as they did, the skipper would yell out in a quick-fire Danish accent, "You betcha—by golly—fer sure!"

Now, at the Beachcomber Bar, I warmed myself with the seductive anesthesia of tequila and mingled with the throngs of fishermen and job hunters around me. The night's talk was filled with affectionate nicknames—names such as Pigpen and Haywire, and Crumley, Jake, Worm and Spiderman. There were Nick the Greek and Alabama Joe, Harpo, and Pogo, as well as Jimbo, Mole, Buster, Snake, Susey Q, Big Hip, and Little Spike. And along with these came the real names of other deckhands—the best of the best—such as Mike Wilson and Ron Eads, Bill Caulkins and Danny O'Malley, Ray Flershinger, Richard Thummel, Rich Wright, Bill Alwert, Billy Williams, Tad Mason, Terry Moore, Beanie Robinson, and the Capri brothers, Dave, Rex, and Tom, Terry Sampson and Larry Garrison, Lance Russel, Dave Hammacker, Don Kyre, and Big Steve Griffing.

Within the confines of this noisy bar, the names of certain fishing vessels took on mythical overtones. Some were huge multimillion dollar vessels stretch-ing up to 170 feet in length. They were said to carry maniacal superman crews who could work for thirty, fifty, even eighty hours without a complaint. Ships such as the *Provider*, the *Rondys*, the *Peggy Joe*, the *Progress*, the *Bering Sea*, the *North Sea* and the *Arctic Sea*; the *Shelikof Strait*, the *Labrador*, the *Bulldog*, the *Enterprise*, the *Rebel*, the *Shaman*, the *Sea Hawk*, the *Juno*, the *Kodiak Queen*, the *Marcy J*, and the *Mitrophania* were just a few of the leading names. They were skippered by men with seawater pumping through their veins—men who, 150 miles to sea, could literally smell the crab. Some such skippers in-cluded John Hall, Harold Jones, Khris Poulsen, Erik Poulsen, Jack Johnson, Harold Daubenspeck, Vern Hall, Wilburn Hall, Teddy "Old Feller" Painter, and now his two sons, Teddy and Gary Painter, Loui Lowenberg, Mike King, Charlie Johnson, and Francis Miller, and the legendary Oscar Dyson.

There was also a smaller class of vessel, often no more than eighty feet in length, whose crews had won the admiration of this tight community and gained a reputation for efficiency and sizable profits. The *Buccaneer* was just one example, as were hard-fishing vessels such as the *Irene-H*, the *Ruff-n-Reddy*, the

Van Elliot, the *Atlantico,* the *Rebel,* the *Amber Dawn,* the *Eagle,* the *Alert,* the *Midnight Sun,* the *Northern King,* the *Mariner,* the *Stevie,* the *Ocean Venture,* the *Cougar,* and even the *Royal Quarry.* And to be a deckhand aboard any of these fine vessels brought with it envy, respect, and status.

Amidst a bombardment of noise some call music, I was working my way through the crowd at the Beachcomber Bar when I came upon Susey. No longer clad in her rain gear, or covered with sloppy chunks of chopped-up herring, or the skin-whitening crust of dried saltwater, she looked strikingly feminine. She was sitting at the bar listening to yet another "cute and cuddly" greenhorn (who had only recently arrived in port from Seattle) brag about a fill-in crab trip he'd lucked into.

Noticing my approach, Susey winked secretly at me. As the greenhorn's dialogue continued, she daintily tilted a cocktail glass to her lips and swilled down half of it. Then, while maintaining an absolutely sincere expression on her face, I heard her ask, "Gee, what's it like out there?" Encouraged by her apparent ignorance about the romantic and dangerous life of a deckhand at sea, the young man bragged on.

Next, I spotted Calhoun, my curly-headed blond crewmate. He was standing on the dance floor hugging what was obviously a close female friend. His eyes were closed and he was smiling through an inebriated grin. They moved at half the speed of the music, and as they lolled from side to side, Calhoun seemed about to fall, yet his hands moved down and over the young lady's shapely backside with a perfect steadiness.

At the end of the bar, I came upon a vacant chair and, while the band took a break, sat and listened to the commotion around me. I could hear the scattered chink and rattle of glasses as they were dispersed and gathered, the piercing shriek of a woman's laughter, and the testosterone croon of several fishermen as they derided a colleague's runaway hyperbole.

At a nearby table, a few veteran sea dogs were drinking together, and I overheard one bearded crewman complain. "Hell, I made forty-seven thousand dollars last year, and I haven't got a goddamned clue as to where it all went!"

Then I saw Calhoun again. With dance partner in hand, he had cornered yet another fine-looking young woman, and as he talked with her, he pointed over at me. Moments later, the young lady appeared from the crowd without waiting to be invited, hopped up into my lap, and began rubbing her hands across my upper torso as if lost in some lewd fantasy. "Is your chest really fifty-five inches?" she cooed.

Suddenly, a carefree hooch-filled Eskimo sporting long black hair, an Indian headband, mukluks, and a black bearskin vest stopped in the middle of the dance floor and screamed. "And there were king crab ten feet wide!" he bragged, holding his arms fish-story wide and repeating it several times.

Some in the crowd laughed at the comical sight. Then, in a friendly tone of mock disgust, one yelled, "Aw, shut up and sit the hell down!"

We had returned to our drinking and had almost forgotten about him when, a few minutes later, the mad hatter in the bearskin vest appeared again.

He staggered out onto the dance floor, leaned back, and bellowed like a bull in heat. "And there was waves a hundred and ten feet high!"

Picking up on the contrary relationship, now several more in the crowd screamed back in a near-perfect unison: "Shut up and sit *down!*"

Then an arm-wrestling match between two crewmen from the same ship flamed suddenly. Table and drinks went flying; but friends on both sides were quick to subdue and separate the men. There was a short and heated exchange of words. Then the two shook hands.

No sooner had they come to terms than our very own "Bear Man" once again stumbled forward from the crowd. Undaunted, and even encouraged by the growing attention, he leaned back like the performer he was and yelled, "And there was sand fleas bigger than *prawns!*"

The response from scores of us was immediate and unanimous. *"Aw, shut the hell up and sit down, will you? fer Christ's sake!"*

SIX

The next morning, I found the *Royal Quarry* tied to the B & B docks and stowed my gear aboard her. Two hours later, Calhoun, Susey, and I were standing on the ship's back deck busily coiling the white corded tie-up lines when Jonesy accelerated out across the sheeny gray waters of Chiniak Bay.

Privacy on board crab boats, I soon realized, was nearly impossible to come by. The tiny stateroom that I shared with Calhoun contained two bunks mounted directly above one another.

At sea, a man's bunk is the only place on board that he can claim as his own. It was nothing more than an oblong cubicle a yard high, a yard wide, and barely six feet in length, but it was a private space, one whose sanctity was generally respected. Mine had a simple yet adequate cot, with one drawer mounted beneath it just inches off the floor. A tiny lamp hung directly overhead, illuminating a small wooden shelf just wide enough to hold *The Grapes of Wrath,* a vitamin bottle, a notepad for my daily entries, and the few letters I had received since arriving in Kodiak.

The aged brown paneling of my surroundings was equally austere and unadorned. There was a pinup calendar, an arrangement of sixteen-penny nails that served as clothes hooks, a porthole that looked out to sea from Calhoun's bunk directly above me, and a tiny two-foot-wide bench wedged into the wall space between the base of my bunk and the wall supporting the door.

The most inconvenient aspect was the ship's latrine and shower. To use either required a trip through the galley and across the back deck to the small steel compartment built into the back of the wheelhouse. In bad weather, the

40

back deck was often flooded with wave wash that cascaded violently from side to side with each roll of the storm-tossed ship. Crossing it, as I well knew, could be an all-out adventure.

The galley was a narrow rectangular space situated between the staterooms and the back deck. It came complete with utensil drawers, a refrigerator anchored to the floor, an oven with a thin steel containment railing surrounding the cooking surfaces, a sink with an eye-level porthole mounted over it, cupboards with snap-lock door latches, a toaster secured to the wall by strands of crab-pot webbing, and a small microwave oven.

Directly adjacent to this stood the galley table, a triangular surface covered with a rubber mat designed to keep dinner plates, beverage glasses, and meals from making disastrous cross-table cruises while at sea. Against the wall, and bordering the dinner table on two sides, were knee-high compartments that served as benches. Beneath the long wooden planks covering these compartments was storage space for canned goods such as soups, vegetables, fruits, and beverages, plus a wide assortment of cereal and essential foodstuffs.

It was blowing hard as we wound our way through the smooth tidal waters of Geese Channel. We crossed Alitak Bay and in a torrent of wind and rain moved up into the headwaters of Dead Man's Bay. Here, late on a lonely gray afternoon, we came upon a good portion of the fleet lying at anchor. With their engines off and the cold winds of winter driving hard against their hulls, their anchor chains extending out almost from their bows were taut with pressure. As the wind-raked ships drifted at anchor, they played back and forth across the lonely gray face of the sea in long arches.

An icy and powerful seventy-knot wind came against us without a lull as Calhoun and I climbed out on the bow. Faced with a chill factor for which there are no charts, we succeeded in dropping our anchor over the side. Satisfied that it had secured an adequate grip on the muddy bottom, we hurried, shivering, back inside the warm protection of our wheelhouse and sipped at the hot cups of cocoa Susey thoughtfully had prepared for us.

Then, climbing into the wheelhouse, the entire crew stood and watched the anchor chain for sign of slippage, inadvertently taking in the bleak-looking stretch of shore and the treeless snow-bound mountains of Kodiak Island rising sharply into the clouds in front of us. When darkness once again enclosed us, I lay in my bunk and listened to the moaning sounds of the wind howling around the cold steel form of our wheelhouse.

Steep and desolate mountains rising up on three sides of our ship offered us protection from the waves, but they had the reverse effect on the wind. As the fierce winds hit the mountains and are channeled into the mountain passes, their force accelerates dramatically, and the mountain passes become virtual wind tunnels. Fishermen working along the Alaska mainland, I was told, have reported williwaw gusts of wind literally exploding down out of the passes at an unimaginable 150 miles per hour.

On this night, occasional gusts began striking the water around us with such force that they actually tore island-sized patches of seawater apart and sent them tumbling across the face of the sea. The wind velocities increased to the point that anchors began losing their grip on the bottom, and ships on either side of us began sliding backward into the lonely black hollow of night. Our marine-band sets were jammed with fishermen cursing the weather, their equipment, and their luck. I now could understand how during the long winter months crab boats often spent long weeks "weathered in," swinging at anchor in the partial protection of some remote bay or inlet.

That night, while Susey was on anchor watch and Calhoun slept, I took time to mend several tears in my pants, sew on a missing shirt button, and close up a few rips in my rain gear with the aid of three-inch-wide strips of silver duct tape. That done, I returned to my bunk and plunged eagerly into a much-shared edition of Jack London's *The Sea Wolf*. Across the hallway, I noticed Jonesy thumbing through a dog-eared, grease-smudged adventure novel called *Sitka,* by Louis L'Amour.

The following day as we lay on anchor, news blared from the VHF radio set that two crewmen from the fishing vessel *Epic* had been washed overboard. It was the kind of nightmare event that all deckhands secretly fear; that is, to become entangled with the line, webbing, or frame of a crab pot and be dragged over the side with it.

Under the best of circumstances, I was told, it was not an easy thing to retrieve a man, but to have it happen during some of the worst Kodiak Island weather of the winter was a nightmare indeed.

Kodiak's Mike Doyle was one of the young crewmen who went overboard. He would later recall for me how, at the time, he'd been working on deck in winds that were gusting to sixty knots. It was miserable weather, with a static air temperature of about twenty degrees Fahrenheit and a chill factor of well below zero.

"People were starting to slide around out on deck pretty good," recalls Doyle, when he crawled halfway inside a crab pot to remove an old bait jar and hang its replacement. It was so cold that, even wearing thick work gloves, his fingers "weren't working too good," he said, and the small but necessary task took far longer than normal.

Then suddenly, the ship's bow plunged down into "an unusually deep trough and a large wave came out of nowhere," he said. It washed Doyle into the crab pot and swept the crab pot over the side.

In an explosion of water, the pot door flipped shut behind him and Doyle found himself trapped inside a 750-pound frame of steel and webbing. It wasn't until he saw bubbles floating past him that he realized he was no longer on deck but was, in fact, sinking straight for the ocean bottom more than five hundred feet below.

If the water pressure of the depths didn't kill him, and hypothermia did not incapacitate him, Doyle knew he might very well drown. As he descended into the depths, he thought to himself, This is it! Instead of panicking, he felt overwhelmed with anger. The pot was tilted at approximately a sixty-degree angle as it descended through the unending fathoms of bitterly cold seawater and dark green space, and Doyle struggled to free himself and find his way out.

He soon discovered that he'd been washed into the back of the crab pot. He pivoted inside its tight triangular confines, then grabbed the strap used to hang the bait jars. Lying upside down, Doyle drew his foot back and kicked open the pot door.

He crawled out the door opening and curled bodily around the end of the pot. The pressure in his ears threatened to explode as he managed to locate the bridle and line leading back toward the buoys and the promise of air and life on the surface above. For the first time, Doyle felt fear race through him, jolting him from head to toe. Then, with head up, he struck out for the surface. He went straight up, kicking and stroking, and as he swam, he soon spied the hull of his ship overhead, framed against the winter gray space of sky. It was bouncing and wallowing in heavy seas perhaps another fifty feet above him.

As he pulled his way through the frightening depths, Doyle could see bubbles swarming around the hull overhead. They could mean only one thing: The *Epic*'s deadly propeller was engaged. He could see the backwash churning out from behind the rotating blades.

As he swam through the deep water, Doyle's lungs began to burn for air and once again he felt a deep sense of anger. He adamantly refused to give in, however, and stroked all the harder toward the surface. There was no time to navigate, and in his urgent need, he moved up and up. The journey ended abruptly when he slammed headfirst into the hull of the ship.

Still submerged, and with no way of knowing which way the skipper would go as he maneuvered into the wind and seas above him, Doyle hesitated. He could hear the high-pitched whir of the propeller and saw horizontal columns of countless bubbles jetting out from it. At one point, his feet were little more than a yard away from the churning prop and flapping rudder, and he could feel the currents pulling at him.

When he finally broke the surface, he gulped in volumes of air. In that same instant, he felt a tremendous pain welling deep in his lungs. Numbed by the incessant cold and exhausted from the effort, he fought to remain afloat.

Then Doyle spied the long, curved hook of a pipe pole (used to snag buoys) being shoved out to him. It was only a few feet in front of him, but he no longer had the strength to reach out for it. Worse, the bitterly cold Gulf of Alaska water had chilled him to the point where he could barely tread water in the choppy seas.

Doyle then spotted Tom Thieson, a crewmate, in the water only ten feet away. He, too, had been swept overboard by the rogue wave, and he had suffered cuts across much of his body when he became entangled in a coil of line that had grown tight during the crab pot's descent.

Doyle could see the frantic movements of his crewmates on the deck of the *Epic* as they tossed buoys and life rings overboard to him. The blustery winds blew each of the rings through the air and off across the face of the sea as fast as they were thrown, however.

"Get me out of this!" he pleaded. But his words were swallowed by the wind, and he watched the *Epic* drift quickly away.

Doyle's hopes plummeted. All the time he had fought his way toward the

43

surface, he felt certain that just to arrive there meant rescue, safety, and comfort. Now this! He knew he had to conserve his energy. Somehow, he would have to keep himself afloat in the energy-sapping water until someone came upon him.

Doyle had been in the water without a life-sustaining survival suit for an incredible twenty minutes when he felt himself start to lose consciousness. He fought against it, and was rewarded by the sight of a buoy floating past him. He grabbed it and hung on.

Doyle could not have known that the *Epic's* skipper had just managed to rescue Tom Thieson and was now maneuvering for another attempt to save Doyle. Out of nowhere, recalls Doyle, the giant form of the ship appeared once again, rising and falling as it came. Helpless to save himself, Doyle could only watch as the steel hull of the ship drifted closer and began to roll over him.

The boat dipped and rocked. Suddenly, Doyle found himself looking into the face of a fellow crewman. The crewman leaned far over the side and, timing his rescue attempt with a particularly steep roll, reached down into the water and grabbed Doyle, bear-hugging him around the chest.

"I've got you, Mike!" the deckhand assured Doyle. "And I'm not going to let you go!"

Mike Doyle felt himself being lifted from the water and dragged over the side and onto the deck. It was the last thing he remembers of the nightmare episode. He'd been submerged in the bitter thirty-nine-degree waters of the Gulf of Alaska for some twenty-five minutes, and without the protective insulation of a survival suit. He had survived the longest recorded immersion without a survival suit in Alaskan history. Delirious from hypothermia, he passed out as soon as he landed on deck. As he lay there, his fellow crewmates rushed to remove his clothes—he was wearing wool longjohns, wool socks, and a wool neck wrap—and tuck him in bed. Doyle did not regain consciousness for three hours.

Back aboard the *Royal Quarry,* we knew little of this. But getting two men washed overboard was no ordinary incident, and as we lay in our bunks waiting for the weather to come down, Calhoun offered a bit of advice.

"It's awful bad weather to be working in right now," he said, lying in his upper bunk and talking to the ceiling. "That's why we're in here. Jonesy won't take any chances. But who knows what or how the *Epic* deal happened. It might have been a freak wave. Or someone might have been throwing buoys over and lifted a foot and got himself snagged. That does happen. But losing two crewmen over the side . . . I don't know. Maybe one got tangled and another man came to help and got caught up in it, too. We'll find out when we get back to Kodiak. Just remember, Spike, when you throw those buoys over, or when you're working near crab-pot lines playing overboard, give them lots of room, and keep both of your feet planted flat on deck. That kind of thing doesn't *have* to happen, you know."

He paused.

I was thinking, going over in my mind all that he had said.

"Spike, you listening?" he asked, his voice slightly irritated.

"What did you say, Calhoun?" I replied. "I didn't catch a word of it."

With the kind of weather we had encountered during my first trip and now this, my second journey that tanner crab season, I was beginning to feel like somewhat of a jinx. Dawn unveiled a much lighter world, for the wind had eased to a workable twenty knots and much of the gloom of the previous day had departed. When we arrived out on the fishing grounds, we immediately began pulling our gear.

The air was thick with tension as the first crab pot arrived on the surface. It carried with it forty-seven tanner crab, a fair catch. The second had sixty-three. Working in six-foot seas, we took full advantage of the decent working conditions and turned over our gear in quick sequence, twelve to fourteen crab pots per hour, hour after hour. We labored through the day and well into the night, moving our poorer fishing pots here and there, as Jonesy saw fit, in search of even better fishing.

Calhoun and Susey took turns working the hydraulics and coiling. It was my job to chop up the twenty-five-pound blocks of frozen herring bait, fill bait jars, and string cod on our stainless-steel hanging bait hooks. In addition, each time a new pot showed on the surface, I ran to hook its bridle and help guide the pot aboard, and to sort, measure, and count the crab that came with it.

It was cold and windy on deck, and though we were constantly inundated with spray, I found the conditions ideal for hard work. During the day, the sleeve end of my cotton sweater slipped out from under my rain-gear jacket. Before I noticed it, the poorly suited cloth had drawn water up the entire sleeve—all the way to the shoulder and beyond—and it quickly chilled me. It was something to remember. Under wet and windy conditions, soaked cotton clothing insulated the body little more effectively than no clothing at all.

The fishing was solid and steady and throughout the day we averaged nearly eighty tanner crab per pot. Calhoun and Susey whistled at the size of the crab as a steady rhythm of three-pound tanner crab clattered melodically against the metal sides of the chute leading into our live holding tank.

Susey was tough and talented, and working alongside her left no room for doubt; Susey really did possess incredible strength for such a young woman. I often thought of the wrestling match I had happened upon in that waterfront bar in Kodiak, and the startled and incredulous look on the face of that unsuspecting male greenhorn.

Susey had tiny crow's-feet, which, when she squinted, sprouted near the corners of her eyes and gave her face a salty, oceangoing appearance. She also had a tan complexion, as if she were part native. And while no one could dispute her strength, her knowledge of the fishing trade, or her willingness to work, at 135 pounds, she was also quite pretty. At sea, she brought a feminine grace to what was commonly a barbaric male existence.

Her hands were matted with penny-sized calluses and, when she chose to, she could coil entire strings of crab pots bare-handed without discomfort. To

45

keep her head dry, she wore a liner stripped out of a hard hat. It was brown and lined with fur and had long earflaps that snapped together beneath her chin. She looked cute and eccentric in it, like a cross between a World War I fighter pilot and a feminine imitation of Snoopy.

"I know it looks dorkie," she told me, "but you shouldn't laugh, because it's comfortable and keeps me warm out here."

To our delight, the productive fishing continued and by the time we had finished running the last of our eighty-six crab pots, we had more than 4,500 live tanner crub stored in our tank.

Then, buffeted by eight-foot seas rolling across ten-foot swells, and faced with rapidly increasing winds, we began the four-hour run back into Lazy Bay to rest and allow our pots time to attract more crab.

Snow was slanting steeply across the water as we idled into the bay. And, as on my first trip, it seemed that the entire fleet had taken refuge there. There was the *City of Seattle,* the *Atlantico,* the *Van Elliot,* and two dozen other crabbers, big and small, swinging on anchor. Here, the crews of the vessels could rest, or make necessary ship repairs as they awaited a favorable change in the weather. By midnight, the wind was howling across Kodiak Island. Even in the cozy, well-protected waters of Lazy Bay, the wind found us, flattening the tiny harbor waves and smearing them into streaks of foam.

Several hours before the dawn of the second day, Susey and I climbed out onto our bow and, standing in the unending gusts of arctic wind, hydraulically winched our anchor and chain aboard. Then Jonesy brought the *Quarry* about and we idled quietly out through the dozing fleet and pounded our way to the fishing grounds. We hauled gear all that day and the next, returning to port only once more during the trip. After we had run through our gear five complete times, our hold became too full to continue and, with nowhere else to store tanner crab, Jonesy turned the *Royal Quarry* toward Kodiak.

We battled our way back to Kodiak against a rough chop and a contrary swell. The trip took eighteen hours. However, when we finally docked in front of the B & B Cannery, Jonesy finished the journey on an upbeat note. He took out his pocket computer and his checkbook and estimated the weight of the load of crab we carried on board (about forty-seven thousand pounds) and multiplied that against the current price for tanner crab. Then he estimated our crew shares, wrote out three checks for one thousand dollars, and handed them out to Susey, Calhoun, and me.

I was studying the zeros on the face of the check when Jonesy spoke to me. "That's as close as I can come to what you have coming without seeing the exact weight totals. But if you need any more during the season, I can give you a draw. Just come and see me."

My spirits were still soaring as I walked out onto our back deck. Then I heard a familiar voice calling down from the dock above me. It was Hungarian Joe. "How did you boys go and do?"

"Hey, Joe!" I yelled back. "We're nearly plugged! What'd you expect? Good God, man!" I hollered back in a tone that implied "Don't we always?"

"Did I not tell you," Joe shouted back happily. "Dis Jonesy, oh, he catch zee crab! Vee unload you now."

We had arrived in port in need of a turbo filter and provisions of food and beverage, so in the morning we scattered to gather them. When the *Royal Quarry* again rolled out of the harbor, it was at "no wake" speed. As we passed the crab boat *Atlantico,* several crewmen on its back deck acknowledged us with friendly and yet indecent gestures.

The night before, I had followed a progression of inebriated escapades to the steps of Bee's whorehouse, a local enterprise popular among more than a few fishermen. Now, after the unexpected solace of the previous night, I felt strangely peaceful. I was single and young, healthy and free. And tied to nothing but the moods of the weather and sea, I soon lost myself in the inescapable personality of the journey at hand.

Bouncing our way across the brilliant chromium silver of Chiniak Bay, we turned south, paralleling the banks of Kodiak Island several miles offshore. It was completely dark when we approached Geese Channel.

Jonesy and the others were asleep, and my orders were to awaken him well before we entered the pass. Yet, now, with two trips worth of wheel-watch time behind me, my confidence soared. Though I had never run it even in daylight, I felt sure I could figure out the navigation markers as we went. It was simple. There was a steering wheel and, up ahead, the buoy lights. Keep to the right of the channel markers. Why, there's nothing to it, I thought to myself. With such a simple piece of navigating ahead, I stood for a long time and contemplated whether to wake Jonesy or take the ship through myself. I knew that the fast-moving waters of Geese Channel called for a serpentine path of meandering long turns. Its tide-swept reefs were marked with the hull-paint scrapings of ships that had wandered, and the decomposing skeletons of those who could not be saved. It was one of my better decisions to awaken Jonesy as I'd been instructed.

Twenty minutes later, as I lay in my bunk, I felt the hull of the *Royal Quarry* strike bottom. "Jesus! We're aground!" I heard Jonesy say aloud.

In only seconds, Susey, Calhoun, and I had rolled out and were standing in the wheelhouse. Jonesy's face was blanched and his hands were shaking. Instinctively, he threw the gears into reverse and opened up the throttle. As the main engine bellowed to life, the hull underfoot began to shake, and Calhoun and I glanced at one another. We knew the hopes of an entire season rested on the outcome of Jonesy's emergency maneuver.

The ship hesitated, then began to grind backward across what sounded like a solid bar of sand. Jonesy winced painfully, as if the ship were a physical extension of himself. And as the *Royal Quarry* picked up momentum, her hull thumped heavily . . . once . . . twice . . . then floated free entirely.

Susey and I let out a collective sigh as Calhoun raced below to check our hull for damage. None was found—nor did we need repairs in the future.

Back in the wheelhouse, I watched Jonesy as he guided the *Royal Quarry* back into the channel. Balanced like a boxer on the balls of his feet, he glanced repeatedly at the map, then back and forth off our bow and stern as he tried to judge our position. Directly off our stern, the green buoy-marker light flashed every three seconds, while ahead of us flashed the next channel marker.

Jonesy turned sideways and raised his arms from his sides, aligning the two markers. It didn't seem possible that we could have gone aground. "I just don't believe it!" he said repeatedly, glancing at the alignment of the channel markers again and again. "We're right on course!" He threw up his arms. *"Damn it!"*

In Lazy Bay, I spoke with him about the incident. Although a number of boats had touched bottom there that season, it was the first mishap of any kind Jonesy had experienced in more than six months of skippering.

"I'm sure glad I got you up when you told me to, Jonesy. I'd been thinking of taking her through myself," I confessed.

"I am, too," he replied. " 'Cause if you *had* been the one to touch bottom like that, I would never have forgiven you."

When we arrived out on the fishing grounds, I plunged into my work with a feeling of gratefulness, knowing that our season, and possibly our lives, could just as well have ended on the reefs of Geese Channel. We ran gear and we stacked gear, and we moved them to other areas and dumped them over the side. And, once again, in little more than a week, we managed to nearly fill our hold with tanner crab.

SEVEN

For the entire season, the *Royal Quarry* was our floating home, our workplace, our vehicle to and from the fishing grounds, and, without excuse, we answered to her day and night.

By the middle of March, I felt comfortable on deck. My seasickness occurred only in the worst of weather now, and the banter I received from Susey and Calhoun came peppered with affectionate tags such as "Master Baitor," "Pot Bull," "Big Guy," and, occasionally, far worse things. By the close of the south-end season, I could tie the essential knots such as the bowline, the clove hitch, and the carrick bend, and coil the line aboard as quickly and with as much control as anyone on board.

On deck, Susey's tireless performance continued. "Always do more than is expected of you," she told me one day. "And if you think there's nothing left to do out here, you just haven't looked hard enough."

With our March calendars promising the arrival of spring, and with the entire fleet's 30-million-pound Kodiak Island tanner crab quota nearly filled, Jonesy courageously decided to venture from the congested waters of the pack. We piled our back deck high with pots and set out on the 125-mile journey west. As we passed the towering cliffs and spires of Cape Ikolik on the southernmost tip of Kodiak Island, I felt vulnerable for the first time aboard the *Royal Quarry*. But without a pause, Jonesy motored past them and headed out across the waters of Shelikof Strait.

We arrived safely on the beautiful and primitive coastline of the Alaska mainland and, with no prior knowledge of where to fish or what to expect, we

49

baited our gear and began dumping it in a half dozen promising locations. Then we returned to the waters off Alitak Bay, gathered up a second load of pots, and hurried back across the dangerous waters of Shelikof Straits once again to check on our luck.

We pulled our first crab pot in the shadow of the bold vertical headland of 2,500-foot-high Cape Kumlik. Susey coiled, and our crab block popped and groaned with a suspicious intensity as the steel trap ascended unseen out of nearly three hundred feet of shoreline sea. As it rose, the silence on deck was deafening, and as the coil of crab line grew on deck, the entire crew, including our skipper, pressed close to the railing in an effort to scan the rich green depths for sign of our pot. It closed suddenly on the surface, appearing as a greenish white block. We held our breaths then, and when it rose from the surface its rectangular frame came loaded with a clinging, pinching, dripping mass of legal-sized tanner crab, perhaps a thousand pounds of the creatures.

"Okay! All right!" I shouted. Calhoun clapped his hands. Susey whistled. Jonesy smiled broadly and climbed back into the wheelhouse.

"We can handle about a hundred more pots with that kind of fishing," Jonesy called out over the loudspeaker. "We'll take that kind of action as long as we can get it! Heck, we'll even put it back. What do you guys say?"

Working with a wonderful enthusiasm, we emptied out the crab, replaced the bait jars, closed and tied the door, tossed the first shot of line on top, and took our positions for launching. After several months of working together, we anticipated one another's moves like people with the powers of telepathy.

"We're ready back here!" Calhoun shouted.

"All right, dump it back over," instructed Jonesy. "No, wait a few seconds till I get away from that other buoy setup. We've got 'em packed in here so tight, there's hardly enough good bottom left to lay a pot!"

Jonesy accelerated ahead and Calhoun braced himself at the hydraulic controls. Susey stood with one hand on the crab pot and a shot of line in the other as they waited for the skipper's go-ahead to launch, I sorted through the mound of crab. It was clean fishing and made for wonderful work, for with almost no females to chuck back over the side, nearly every crab in the writhing pile was large and legal and fell heavily into our live-tank chute.

As I worked, I heard Jonesy's launch order. Then came the hissing whine of the hydraulic arm lifting the pot, followed by a metallic clatter and a foamy ruckus as the pot slid from the tilting rack and splashed overboard into the passing sea. Susey tossed two more successive shots of line over the side, then stooped and gathered the buoy setup and tossed it overboard. Each of the three buoys in the setup arched high and slapped the surface with a resounding *kwack*, landing well away from the snagging prop and rudder on our stern.

"How many did we have in that one?" Jonesy called as he began slowing for the next pot. Susey and I gave Calhoun our individual counts. Adding the three together in his head, he turned and spoke into the two-way speaker mounted on the back wall of the wheelhouse. "There were two hundred and twenty-five in that pot, Jonesy," he announced happily, unable to hide the pride in his voice.

50

Five days and as many turns through the gear later, with our tank over-flowing with tanner crab and our crab pots rebaited and fishing, Jonesy pointed the bow of the *Royal Quarry* toward the island of Kodiak and we headed for home, where we would off-load, resupply, and return to our pots at the first opportunity.

Jonesy, we soon realized, had brought us to the raw edge of unaffected earth. A more magnificently wild and untouched country I could not have imagined. Here, a man could take joy in his work. Here, a man could lay to rest the past and, laboring silently for weeks, withdraw for a time from the irreverent and prying eye of civilized man.

Beginning with the bold headlands of Cape Kumlik rising some 2,500 feet above the sea, we traveled northeast along the coast, dumping our pots near Aniakchak Bay, Amber Bay, and Yantarni Bay, and, farther on, the bays of Nakalilok, Chiginagak, Port Wrangell, and beyond.

Prospecting along this seventy-mile stretch of coastline, we labored in the shadow of Mt. Aniakchak and the unbroken glacier-filled horizon of the Alaska Range. We passed between the familiar island outcroppings of Kumlik and Hydra, of Ugaiushak and Navy, and the tanned pillarlike rocks of the Aiugnak Columns. I often felt dwarfed and intimidated by the unmatched dimensions of such a land.

Each day as we turned over our gear, we delighted in the teeming popula-tions of tanner crab that filled our gear, sometimes after only a single overnight soak. Not a day passed without my knowing the satisfaction of a job well done, or the adrenaline rush of greed or danger, or a pristine vision of wilderness, sustaining and unforgettable.

Often as we traveled through those wild and uncharted waters, we spotted brown bears foraging along the shore, moose feeding in the lowlands, and caribou grazing in the valleys or moving across the steep hillsides that fell off abruptly into the sea. Once, as we powered inside Kumlik Island, a flighty herd of perhaps seventy of the grayish black deer came galloping over a plateau. Spooked by a wolf pack, spring fever, or the canyon echo of our diesel engine, they seemed to flow across the land like a living stream. And as they moved, the surefooted creatures skirted the sheer rock edge of a precipice which, even from the safety of our ship, appeared deadly and dizzying. Sea otters proliferated in the area, and some days we counted as many as sixty of the furry creatures bobbing playfully in the kelp. Occasionally, we came upon a solitary humpback whale feeding in the area.

Many of the spectacular rock formations jutting skyward from the water had been smoothed and thinned near the waterline by the forces of time and surf, and their granite faces were bleached white with the droppings of countless swarms of seabirds that looped to and from the precarious perches they had built upon them.

For weeks, we lived a misanthropic lifestyle, roaming along hundreds of miles of magnificent and uninhabited coastline, rarely spying another ship. The majestic land was all ours, and in the wheelhouse of the *Royal Quarry*, Jonesy made decisions and altered course just as he saw fit, much the way a Rocky Mountain fur trapper in the 1820s might have done, turning his mount this way or that, trying his luck in a place he had never before tested, never before seen.

Through the winter season, this mainland country lay frozen beyond cold by incessant winds and blowing snow. Now, in March, the calendar proclaimed it spring, but the promise of warmth remained unfulfilled. Everywhere we traveled along the wintery coast, blustery winds cold enough to crack teeth rushed down out of the glacial passes and struck us with the crude insensitivity of a drunken brawl.

April seemed to tame the winds and sea. Then for entire days, the mood of those mainland waters lingered as calm and gentle as a candlelight dinner. Day after day, the *Quarry* slid across a plate of sea so smooth one could almost feel it, like the boyhood idling of a thumb across the polished surface of an agate.

Whenever the fickle Alaskan sun found me on deck, the tough vinyl material of my rain gear soon hung limply around me, growing as soft and pliant as the pocket leather of a catcher's mitt. On those days, both at dawn and again at dusk, the water took on the natural greenish blue luster of a salmon's back. And if, while pulling crab pots in the cloudless light of midday, you glanced over the side, you could see emerald green prisms of light playing off into the depths. Like the glimmer of a sapphire star, the colorful display would parallel your every move.

We had it all that spring: youth and ambition, freedom and luck in a wild and scarless land. As we worked our way along the coastline, we ran our gear by daylight, and often paused for a day to relax, beach-comb, dig clams, or go hunting. There was something comforting about making a living on ocean water so near to such a majestic land and anchoring up each night in the womblike security of a bay, surrounded and protected by it.

It was unusual to have such a long stretch of prime crabbing country to ourselves. Now and then as we worked our way southwest along the coast and drew near to Chignik Bay, we would come upon a limit seiner (salmon boat) working. Outfitted with a hydraulic winch block and a dozen or so small crab pots, these relatively small boats did not have the pumping capabilities of ships outfitted for crab. Unable to keep their catch alive for long, they were forced to return each night to the tiny Aleut village of Chignik to off-load.

Occasionally, the dominant outline of a Bering Sea crab boat loomed ominously on the horizon, but none lingered for long. They stretched for more than twice our length across the water, and were understandably uncomfortable maneuvering in such close quarters.

I had one especially frightful experience that season. It happened on a return trip to Kodiak, after another fine and profitable voyage to the mainland.

The tides had been running strong through the waters of Shelikof Strait, and the wind was blowing contrary to the tide, a condition that usually produces steep, lapping waves. Yet it was not so much the height of these waves that unsettled fishermen, it was the unusually short three- and four-second intervals between the departure of one wave and the arrival of the next. Crab-boat skippers trying to move through such seas found that after rebounding from the thrust of one wave, there was too little time to recover before the next wave would catch them and collapse onto their bows, crash over their wheelhouses, or, worst of all, explode through their wheelhouse windows.

I knew that the windblown Shelikof Strait that stretched for a good one hundred miles between Kodiak Island and the Alaska mainland had a nasty reputation. With dependable regularity, her steep combers punched in wheelhouse windows, tore antenna equipment and radar gear from wheelhouse roofs, and washed crewmen working on deck off their feet, while her fickle currents sucked crab-pot buoys under and sometimes kept them submerged there, useless to their owners, for weeks at a time.

The waves in Shelikof Strait were rough but negotiable the night Susey called me for my watch, one of scores of watches I'd taken that season. Pounding into such steep close-crowned seas made for the worst kind of bucking ride, and I'd been seasick since we left the shelter of the mainland. Now, my entire body felt as if it had been flogged with a lead-filled length of riot hose as I listened to Susey's debriefing. "From what I can see, there's no one near us," she began as she fought with the chest-high wooden steering wheel. "But do you see that?" she added, pointing to a foggy island of green outlined with each revolving sweep of the radar screen's searching arm. "That's a rain squall. It's about an hour away. I don't know what could be inside that; probably nothing, but . . ." Suddenly, spray smacked heavily against the window in front of her and Susey tossed the wide wooden steering wheel three spokes to port. "Now, if you have any problem, or if you have any questions, you know the rules; wake the skipper. Or come get me up."

As she spoke, I felt the irrepressible green rush of nausea coming over me. I staggered to the starboard side of the wheelhouse, opened the top half of the Dutch door, and, leaning out into the wet spray and wind, retched myself empty.

Then I slammed the door shut, wiped my face dry, and took over. "Jesus, you look like death! You going to be okay?" Susey asked me, chuckling. "Spike," she continued, "if I got as sick as you do out here, I'd never hire on in the first place. How the hell do you stand it, puking up all the time and looking like you're snake-bit and about to die?"

The *Royal Quarry* was falling off to port then, so I threw the wheel half a revolution to starboard to correct the drift. "Susey, when I'm out here, I swear to God that I'll never return. It's too miserable, I tell myself. It isn't worth it. To hell with the money. But when the trip's over, I can't seem to recall how bad it really was, and I stay on for another trip. I guess I hate being broke worse than I hate being seasick."

"I think you're sick, all right," she said finally. "You got her now?" she asked, pausing in the stairwell. "We're taking two-and-a-half-hour watches. Yours will be over at 4:30 A.M. Calhoun's up next. But Jonesy may decide to relieve you before that. He does that sometimes."

Susey went below then, and over the next several hours, as we neared the

53

growing island of fog and rain, I glanced periodically at the radar screen. Something didn't seem quite right. Inside the squall was an area that seemed slightly more substantial than the green haze surrounding it. There was no way to know for sure. I contemplated waking the skipper but decided against it, and remaining true to our course, I nosed the *Royal Quarry* into the heart of the turbulence.

Then, without a moment's notice, the wheelhouse windows flooded with rain, melting much of my vision. The radar screen clouded completely, and I began to have strong misgivings about the situation. Susey had warned me about taking on too much. "If you've got any doubts while on wheel watch, never be too proud to get someone else up. Going it alone can sink ships and kill people."

Jonesy had been absolutely adamant. "If you've got any questions, if you feel yourself getting into a predicament, or are faced with something you just don't understand, then for God's sake, get me up! I won't mind it if you do. I'll be good and mad if you don't!"

But somehow, the combination of embarrassment at not being able to hold up my end of the work load and the hope that the present situation would soon resolve itself won out, and I journeyed foolishly on.

As I moved ahead through the pitch-black collusion of rain and fog, our mast lights could reach no farther than a single wave beyond our bow. I felt almost too weak to stand, and I found myself squinting through a dizzying wash of rain melting against our windows, fighting to meet each new jerk of the storm-tossed wheel. Unwilling to call for help and unable to leave my duties to vomit out the back door, I kept a towel close at hand and periodically puked into it as I struggled to maintain our original compass heading.

Without a hint of the coming change, the *Quarry* moved out of the rain, and I found myself powering blindly ahead through a peninsula of boiling fog perhaps fifty feet deep. Then I broke free of the fog, but when I looked up from the compass, there, illuminated in detail in the direct beams of our mast lights, appeared the towering hulk of a processing ship. It was steaming broadside across our bow, dead ahead.

The ship had come out of nowhere, materializing before me like a lethal hallucination. There seemed no way to avoid striking her, and as her midships slid by, I stored the inadvertent flash of a ghostly vision off our port side. It was the gray-black outline of the ship's stern slipping from the fog bank. Long tendrils of steam trailed after her.

The giant ship stretched for perhaps five hundred feet across the water, and we were about to plow dead center into her midships. Frantic with fear, I threw the wheel to starboard as far and fast as it would spin and heard myself praying aloud, superstitiously and without thought or hope.

"Oh-God-oh-God-oh-God-oh-God" was all that came out. With the wheel locked full over, the *Quarry* hesitated, then began to swing to starboard. As it inched around, I looked up and saw the gray elephant skin of the massive ship. It passed parallel to us as we turned, sliding by so close to us that it seemed I could reach out and touch it. Then, overhead, I spied the yellow circles of portholes and, in them, faces staring out at me and hands cupped against the light of their rooms. Then a series of angry staccato blasts shook the night and the screams of an irate helmsman bellowed from our radio set, and I rushed to silence the volume.

I steered the *Quarry* on around, cutting across the foamy path of our own wake. We rocked sharply as we swung broadside through the deep ocean trough. When we had completed a 360-degree circle, I tossed the wheel hard to port and returned us to our original course. Then I stood frozen at the wheel and I listened for the footsteps of others, but heard none, and never mentioned the incident to anyone.

Of course, it was no one's fault but my own. Even the best of skippering men could not remain afoot and hover over their men forever. Sooner or later, a skipper had to grab a little sleep and trust a few hours of steering to his crew. Jonesy had been up for more than thirty hours. I'd failed him bitterly, as well as my crewmates, and my self-loathing went beyond shame.

EIGHT

In early May, our season ended. According to the terms I had agreed to with Mike Jones when I first hired on to the *Royal Quarry*, when the season expired so would my job. The position I had filled temporarily, I understood from the beginning, had been promised to a scrappy young wrestling buddy of Jonesy's named Tim Gerding. He was just now ending the year's classes at Oregon State University. Jonesy had carefully explained all this to me when I first came aboard, and I had no problem with it. I felt only gratitude toward him for the entire experience, as well as the many friendships I'd made within the fleet during the season.

Susey and Calhoun would remain with the *Royal Quarry*. Throughout the summer, they would transport boatloads of salmon from the fishermen working out on the grounds to the B & B Cannery in Kodiak. While they went on, secure in their hard-won positions aboard the *Quarry*, I held out one hope, and that was to land another crab-boat job in time for the high-stakes bread-and-butter king crab season that would kick off in the fall.

The entire Kodiak tanner crab fleet (composed of some 135 boats) had caught and delivered more than 30 million pounds of tanner crab to the Kodiak canneries during the three-month tanner crab season we had just completed. In the past month, the 121-foot *Rondys,* for example, had delivered 240,000 pounds of tanner crab in a single trip, and skipper Vern Hall and his crew had "put in" (delivered) an incredible 2 million pounds that season. Yet, while he and several other huge floating steel legends (such as the 120-foot *Provider*) delivered such huge amounts of crab, other crab boats, both small and large, struggled to top 250,000 pounds for the whole season.

In his first tanner crab season ever, Mike Jones, fishing in a seventy-one-foot crab boat with a live-tank carrying capacity of only forty-nine thousand pounds per trip, had managed to catch and deliver 560,000 pounds of crab, a total that placed him near the elite, ultracompetitive top of the midsized crab boats in the entire American fleet.

As agreed, when it came time to return the leased crab pots, we craned them ashore. Before we stacked them for storage, however, Jonesy, Calhoun, Susey, and I worked to repair them. We mended every webbing tear and replaced every door tie, chafed bridle, shot of line, and lost buoy with new ones. Typical of the way our youthful skipper ran things, when the crab pots were finally stacked in neat rows ashore, they were in far better condition than when we had first leased them.

"Well, I guess that's it, Spike," said Jonesy after the last pot had been repaired and pushed to the end of the last long neat row of pots. He shook my hand. "We'll be running into each other. You did a good job for us. You can be proud of that."

"Thanks for everything, Jonesy," I said, feeling the awkwardness of the moment.

Jonesy was right. I *was* proud. But more than that, I was grateful for the experience and fellowship at sea and for the apprenticeship he'd given me. I would never forget the upright, straightforward way he had treated his crew.

As Susey put it, I'd become an "old hand" on deck. I could coil crab line as fast as our hydraulic block could reel it aboard, and my knot tying had improved to where I could tie the all-essential carrick-bend knot on a leaping, wave-washed deck in about six seconds. I'd learned to splice line, stack gear, mend web, measure crab, cook in rough seas, and live closely with others in tight, claustrophobic quarters. I'd been patiently taught to navigate, to use the radar, and to read maps. I knew, and was known to, more than a dozen local crab-boat skippers. And I'd pocketed more than thirteen thousand dollars in the process.

"If I were you," Susey suggested in parting, "I'd try for a job on one of those big Bering Sea boats. It is more dangerous. I mean, they do work out on the edge. But that's where the real money is. You wouldn't believe the kind of money those crewmen are making! I heard that the tanner crab fleet out of Dutch Harbor has put in about eighty million pounds this season. Can you believe it? Some of those guys are pulling down forty-thousand-dollar shares! And that's just for tanner crab. During the king crab season in the fall, they'll make twice that!" She paused, looked down, and shook her head.

"But I won't kid myself. I know I'm too small for that kind of fishing. They work with those big seven- and eight-foot crab pots. Why, the decks alone on some of those ships are one hundred and ten feet long. That's more than forty feet longer than the entire length of the *Royal Quarry* . . . bow, wheelhouse, and all! But, Spike, you could do it. You've got the size and strength . . . if you don't die from seasickness first."

I stepped forward and bear-hugged her.

"Now, just don't forget where you came from," she added nostalgically as she buried her head against my chest and squeezed back. Then, freeing herself suddenly from my grip, she hiked silently back across the dock to the *Royal Quarry* and climbed aboard.

We were a characteristically youthful crew. Susey Wagner was twenty years old, while Calhoun and I were twenty-seven. Jonesy, our skipper, was the "old man" at thirty. Superb crewman and close friend Steve Calhoun would remain with Jonesy and the *Royal Quarry* and go on to become a respected skipper and, eventually, part owner of the boat.

Mike Jones had repeatedly proven that he possessed intelligence, tenacity, verve, and instinct—qualities that were inherent, not taught. Over the next few years, his reputation would grow until he would be generally recognized as one of the finest fishermen for midsized crab boats in all of Alaska.

As for Susey, she would go on to make more than forty thousand dollars aboard the *Royal Quarry* in the next year alone. She bought her own house in Kodiak and in between seasons kept herself on the run. She tended bar at the B & B and hustled on the side as a commercial diver. She inspected ship hulls and replaced zinc plates melted by electrolysis. She was often called on to dive down and cut gillnets and seine nets and crab-pot lines wrapped around the propellers and rudders of crab boats and salmon seiners. When Mike Jones eventually built another crab boat, he named it *Sweet Sea*. He hired Susey Wagner to skipper it.

PART
TWO

THE BERING SEA

SEASON:

KING CRAB FISHING

ABOARD THE *WILLIWAW*

WIND

ONE

I met my new skipper entirely by chance during the summer of 1978. Our paths first crossed in a bar on the waterfront in Ketchikan, my favorite of all Alaskan villages. I bought him drinks through a long and stormy night, and parted by giving him a scrap of paper with my name and phone number written across its face. Lars Hildemar was his name. He was tall and lean, and bald and grumpy; the sixty-one-year-old Norwegian skipper of the *Williwaw Wind,* one of the largest of Bering Sea king crab boats.* At the time, he had seemed impressed by my size and by the fact that for several years I'd worked as a logger and commercial diver. But he'd made it clear that he had a full crew and could promise nothing.

The next morning I was there to see him and his crew off, and I raced back and forth along the dock to throw off his ship's two-inch hawser lines. I was able to make a clean job of it, flipping the lines over the ship's railing and watching with pride as, one by one, the lines slapped down onto the deck next to the waiting crewman, without touching the water or the man. It was a way of sparing them the bare-knuckled, hand-numbing chore of dragging each hawser line aboard through the salt water.

"Keep your nose into the wind!" I yelled, waving them off on their journey south to Seattle. I stood and watched them until I saw the huge black steel vessel turn the corner and disappear from view around the south end of Pennock

* The name of this ship and the names of her captain and crew have been changed.

Island. I was struck by an urgent longing to be shoving off with them, to share their adventures and once again become part of a world that was larger and more important than myself.

I had rented a room in the Union Rooms Hotel, only a few paces from the rich ocean water of Tongass Narrows. Shirley McAllister, the hotel's owner and manager, was a good-looking robust woman with the biggest heart in all of Ketchikan. When I returned to the hotel, Shirley noticed my serious expression. "Your ship's gonna come in, kid," she said, inviting me into her living room for a drink. "Lots of crab boats headed to Kodiak and Dutch Harbor come right through here. Hell, you've got just as good a chance of landing a job here as anywhere. Just spend as much time down on the docks as you can. Be patient and keep looking. Things have a way of working out." She paused and studied me. "You can live here free until you find work," she said finally. "You hungry?"

Several weeks later, I was lying in the soft bed in my waterfront room when someone pounded anxiously on my door. Then I heard Shirley call out to me, "Spike, you've got a phone call! There's a guy down on the lobby phone who says he's calling from Seattle! He says he wants to talk with you about a fishing job." I threw on a shirt and a pair of pants and fled barefooted past her down the hotel stairs.

"You want to come and go to vork for me?" asked a voice flavored with a Norwegian accent.

Thinking it a prank of one of my friends, I exploded into the phone. "Who the hell wants to know?"

"This is Lars Hildemar," he replied, sounding a little taken aback. "I thought you wanted to go king crab fishing. Well, a spot has just come open on my ship. I had to let a man go, you see."

It took me a moment to recognize the voice and catch my bearings. It was the skipper of the *Williwaw Wind* and he was offering me a full-share deckhand berth on his ship during the Bering Sea king crab season in September. His crew was rigging up several hundred king crab pots down in Seattle.

"How soon do you want me down there, Lars?" I said.

"Yesterday would be good," he replied, only half-joking.

The next day, I arrived by plane in Seattle. That afternoon, I joined an experienced crew of older Norwegian crewmen rigging up crab pots on the westernmost edge of Lake Washington on the Ballard waterfront. All were well into their forties, and one past fifty, and though they seemed ancient compared to the deckhands I'd known, they quickly demonstrated that they knew their trade.

The ship's engineer was a friendly, tooth-decayed crewman who had developed quite a reputation for doing what he did, and doing it well. He enjoyed striking up whatever conversation he could, as long as the talk had to do with welding, gears, hydraulics, or diesel engines "bigger than goddamned dump trucks!" He talked of repairs made in high-running seas, "with nothin' but a come-along, a crescent wrench, and balls bigger than a bull's" to see him through.

He claimed to be flat broke. But when Lars offered him a choice of three thousand a month in wages or a 3 percent share of whatever we caught, he laughed aloud.

"Oh, no!" he began. "I've had my share of those percentage deals! I want my wages up front where I can see them! You guys catch the crab, and I'll keep this oversized bucket running as long as the season lasts!" Lars quickly agreed. The mechanic's decision would eventually cost him $25,000.

One of my deckmates was Fritz Boettner. Half Aleut (pronounced al-ee-oot) and half German, he spoke with an accent all his own. He rarely referred to his past life, but aboard ship he was a pleasure to be around. Ruggedly built, he was quiet, considerate, and easygoing, and kept mostly to himself. I was told that whether he slept at home or aboard the boat, he always kept a loaded Browning automatic 9mm pistol stuffed beneath his pillow.

Forty-year-old Sigmund Holten (referred to as "Sig") was another of my Norwegian crewmates. He'd worked for years aboard tugboats that ran between Seattle and the far reaches of Alaska. When he first went to work aboard a Foss tugboat, he soon discovered that if he scrimped and saved, it was possible to tuck away about a thousand dollars a month. And so he held tightly to the job and to every penny, and eight years (ninety-six months) later he had saved more than ninety thousand dollars, after taxes.

Finally, there was our deck boss, Robert "Bobby" Ragdé (pronounced "bob-bee" by our skipper). It was said that with more than forty years of sea time under his belt, Bobby could gaze into the ship's pale green radar screen nearly anywhere along the seven hundred miles of wilderness shore that stretched between Kodiak and Dutch Harbor and, at a glance, give a reliable estimate as to the ship's position.

Now in his mid-fifties, Bobby had been fishing since he was a child. At the age of twelve, he began working a cod line from an open dory off the Norwegian coast in the North Sea. Now when he told stories and looked back, his mind could shuffle through whole decades of past adventures and seasons.

Bobby was quite small for a deckhand—tiny, actually. Working on deck or in the pot yards, he wore an ancient baseball cap that had been coated fore and aft with several years' worth of dirt and grease. It sat on his head in a limp, squatty puddle. His face was weathered from the decades spent exposed to the elements of open decks. And whenever he pitted his thin frame against the impossible 750-pound tonnage of one of our crab pots, the brown skin of his forehead would wrinkle up like the skin of a codfish drying in the sun.

Bobby seemed older than his fifty-four years—except for his eyes, which were icy blue and tinted with a silvery gray pigment. They sparkled mischievously out from under the bushy clumps of his eyebrows, and when we were working, they missed nothing. His hands were small and scarred from a lifetime of work. Sore, I suspected, with arthritis, he would sit for hours at the galley table and rub one hand into the other and tell stories out of the past that were laced with an acrid bit of humor, or joke with his countrymen in his native tongue.

Noticeable, too, was Bobby's angular way of walking, which I assumed was more the result of injury than of style. I would later learn that during one crab season, now more than a decade gone, a crab pot had smashed into his left shoulder. The incident had left a large atrophied hollow in his deltoid and pectoral muscles where the bone and muscle had been crushed and never fully recovered.

While Bobby was smaller and older than the average deckhand, he was intelligent and clever and more than compensated for the strength, stamina, and resiliency of much younger men by running the hydraulic controls on both the deck crane and the pot launcher, and by organizing the methods of our work on deck with a precision one could acquire only through decades of experience.

The third week in August was our deadline, and from the beginning, the Norwegians set a furious pace. Preparing each crab pot for the season ahead, we wrenched on bridle knots, spliced eyes in door ties, tightened runway webbing, replaced damaged doors, and measured and cut and coiled thirty-three-fathom shots of polyester line, melting each end of each shot of line with an electric "hot knife." We pumped the giant fluorescent orange crab-pot buoys full of air, stenciled them with our boat sign—the black foot-high letters *WW*—and, using a self-tightening hangman's knot, tied two of them together with one unsinkable foam "sea lion" buoy to form our own personalized buoy setup.

Like most deckhand chores, the work was repetitive and continued, one task into the next, almost without pause. We worked seven days a week to prepare our gear for the day when we could load up, fuel up, and be gone. Each day as we worked on the banks of Lake Washington and the summer fled toward its heated end, pot-encumbered crab boats would pass us by on the first, idling leg of the sixteen-hundred-mile journey north to Alaska's Bering Sea.

We put in long days on the gear, and during the nights I shuffled tiredly into my rented studio apartment in Ballard to shower and change clothes. Then I would either eat out at a fast-food restaurant or prepare myself a meal of beans and hot dogs, or Hamburger Helper, or inhale broad platefuls of steak and eggs, with baked potatoes, 2 percent milk, and wheat bread smothered with real butter—perennial staples of a contented bachelor.

Each evening, I chose between several hours of reading, a visit to a nearby gym for a quick workout, or a stroll up Ballard Avenue for a shot of tequila and a squirt of lemon. There, uptown, I might stumble upon my Norwegian crewmates at any one of the local watering holes, such as the Plantation Bar, Hattie's Hat, Vasa Sea, Lock Spot, or Shilshole Broiler. These men were an independent lot. They worked hard and they drank hard and, like myself, seemed to appreciate the solitude and peace of mind of being single and free.

It was not until late August that I heard our huge diesel engine grind over and belch quickly to life. No one came to see us off. The entire crew had chosen to say goodbye to family and friends the night before. And so, with a twenty-five-foot-high pile of 240 crab pots (the *Royal Quarry* packed only 38) covering our deck from wheelhouse to fo'c'sle, it was our turn to head north. We idled proudly out along the waterfront, under the Ballard Bridge, past the Foss Shipyards with her seagoing tugboats, past the marinas, and businesses and streets clogged with the lonely rush of traffic.

As we neared the Ballard Locks, Lars called out on the deck loudspeaker. "Big fella, you go up on the forward deck there and work the bowline." It was a command that would stand throughout the entire season. "We vill need two of them," he added.

I had learned much about docking during my season aboard the *Royal Quarry.* The hawser lines and the logistics of springing the entire ship off a properly secured bowline were the same on most ships, but here the size of the lines and the squealing tonnage of pressure on them had increased to almost mind-boggling proportions.

As we entered the lock, my heart raced with worry, but Lars walked me through it with consideration and without incident. "Make the bowline fast right there now, big fella," he said. "All right now, big fella, give me a little more slack on the bowline. Good. Now make her fast."

The ship was still moving forward, so I hurriedly bent and gave the thick hawser line two complete wraps, weaving it in a horizontal figure-eight around the two-foot iron cleat. Then I gripped the end of the hawser line and leaned back and pulled hard against it to bind the knot. In spite of the fact that he had begun reversing his engines, the *Williwaw Wind* continued its forward slide. The line groaned loudly and made little explosions as the immense weight of our ship came to play against the knot. However, the knot held fast as the three-inch hawser tightened around the rounded thick steel of the huge cleat.

There is a twelve-foot drop between Lake Washington and the waters of Puget Sound. With lines leading to shore from both sides of our bow and stern, the massive door closed behind us, and the giant lock pumps kicked in and began lowering us down to the level of the waters of Puget Sound.

On our starboard side, a park with a close-clipped lawn bordered the area. As ships come and go, people stroll past them only yards away. We were being processed through when a small, sharply dressed older man walking a golden cocker spaniel strolled near and paused to look over the *Williwaw Wind.*

"Hello," I said, feeling his gaze from the sidewalk slightly below me and perhaps twenty feet away.

"Where are you headed?" he asked. There was kindness in his voice. I liked the way he treated his dog.

"Alaska. The Bering Sea," I said. Acting as if I knew where I was going, I rested a foot on a cleat and, leaning against the bow railing, glanced down on him.

"And what do you catch in those traps?" he asked, pointing to the massive pile of king crab gear.

"We call them pots. And we're going north to give the fall king crab season our best shot."

"What do those crab pots weigh?"

"Seven hundred and fifty pounds each, absolutely empty," I bragged.

"What would happen if one of them was to fall on you?" he asked finally.

"It would hurt," I answered, nodding.

The man smiled and when he spoke again, longing filled his voice.

"I've never been there. I wish I was going with you."

"Well, to tell you the truth, I haven't been quite that far north, either. If it was up to me, you could hop aboard and go with us. I understand it takes nine days to get to Dutch Harbor, if the seas aren't running too high, traveling day and night."

The man chuckled as he contemplated the thought. "Oh, I'm too old." Then, changing his tack, he began to stroll away. Farther down the sidewalk, the small man paused and turned, and this time his voice was filled with gravity.

"I do know it's a big country up there. A lot of things can happen. You be real careful. And good luck."

I knew what he meant. Anything could happen. And it could come from any direction: the storms, the work, the sea, bad luck, a barroom scrape, an uncharted reef, a blown engine, or a disgruntled skipper. But as we idled from the lock and out into Puget Sound, the fears I secretly harbored were subdued by the enticing possibility of fortune and adventure in Dutch Harbor and the Bering Sea, a part of the Alaskan wilderness I had never seen before.

TWO

Night and day, at better than ten knots (and 250 miles a day), we made our way toward Alaska. It was dark when, several days later, we broke out of the calm, heavily timbered protection of the inside passage between the Queen Charlotte Islands and Canada's mainland shore and, swinging to port, passed through the Canadian side of Dixon's Entrance and moved out into the waters of the Gulf of Alaska. We cleared Cape Fox and followed the "great circle" navigational route to Kodiak Island, encountering moderate winds and seas and favorable running weather across the six hundred miles or so of Gulf of Alaska water.

We stopped briefly in Kodiak to off-load several skiffs we'd transported from Seattle as a favor for a cannery, then exited by way of Chiniak Bay. We turned sharply to starboard and, passing inside Ugak Island, moved parallel to the shore of Kodiak Island.

Dawn found us cruising across a dark oily-blue sea in the lee of an island of steep emerald green hills that stretched without bounds toward the starboard horizon. Bathed in a bright summer green and covered with a junglelike growth of lush green vegetation, the huge island looked reborn.

Throughout the previous winter I'd spent aboard the *Royal Quarry*, the hills and valleys of Kodiak Island had always looked lonely, barren, and even foreboding. But now the steep ruggedness of the slopes and the lush green beauty of the unmarred valleys left me feeling transported. It was as if I was paralleling the shores of Kona or Molokai in the Hawaiian Islands.

Less than a day later, we passed through the Geese Island channel. Pointing our bow toward the Alaskan mainland almost due west of us, we moved past the tall spires of rock at Cape Ikolik at the southernmost tip of Kodiak Island and out across the rocky waters of Shelikof Strait.

Soon, we were closing on the Alaskan mainland near the mouth of Chignik Bay, and the captivating geological formations at Castle Cape. I had seen them first in a Sidney Lawrence painting—the brown pillars of eroding sandstone standing like spirit figures above the entrance to the bay, and a late-nineteenth-century sailing ship, her sails filled with unbridled wind, frozen on the canvas in vivid greens and browns as she cut across a rough and romantic sea.

Now, however, the winds were calm, and the light subdued, and the brown sandstone pinnacles rose up and disappeared into a soft gray fog. On board the *Williwaw Wind,* there was no time for historic fact or artistic allusion. We were closing on the Bering Sea by the shortest, fastest route possible, and we were slicing a course through short choppy seas at a credible eleven-and-a-half-knot pace with the engine throttle shoved in the "all ahead full" position.

Regardless of the urgency of our passing, the shore of the Alaskan mainland was not something to be ignored. It is an immense and rugged country, and it rose and fell beside us for hundreds of miles in wild and scenic accents of rock, glacier, volcano, valley, and plateau. All along the shoreline, rock cliffs leapt skyward from the icy north Pacific in sheer and vertical angles. Standing bolt upright against the wind and waves, they were at once magnificent and frightening. And near Cape Kupreanof, they rose from the sea without beach or warning, soaring straight up for more than a thousand feet.

Day and night, along the way, we passed countless numbers of the tropical-looking puffin birds and hundreds of sea otters. Up in the valleys, I was assured, moose and caribou and grizzly bears proliferated.

Then we passed the village of Sand Point off our port side, and King Cove off our starboard, and as we neared the end of the Alaska Peninsula and the shortcut route to the Bering Sea known as False Pass, the island of Unimak came into view.

Studded with volcanic peaks, both active and dead, Unimak Island is seventy-five miles long and twenty miles wide. The first island in the Aleutian Islands chain, it stretched across our charts in the broad angular shape of a peanut.

By cutting through False Pass, a crab boat could bypass the circumnavigation of the island altogether and move straight across the tip of the Alaska mainland and out into the Bering Sea, a savings of the better part of a day of precious running time each way.

But there was a catch. Crab boats that wandered out of the poorly marked channel that winds through the mud flats of False Pass were striking bottom with increasing frequency. The main channel was not only shallow but also narrow, and at night it was nearly impossible for large crab boats to pick their way through the dark and windy channel. Even when traveling in the full light of day, the buoy markers were no more visible than rusty five-gallon gas cans. Nevertheless, we were running late, and with the season already under way, Lars committed us to the journey without debate.

We powered our way through the flat, swirling currents of Whirl Point, past the steep grassy shores bordering the narrow Gulf of Alaska side of Isanotski Strait, and moved into the flat windswept waters of the pass.

Shortly, we came upon the village of False Pass huddled at the base of a range of mountains off to our port side. In front of the settlement, a pier extended out from the shore. Its wooden planked deck was covered with the green monofilament piles of gillnets, and the black mesh piles of seiner nets, their white buoys gleaming in the late-afternoon light.

The pier stretched back from deep docking water to a cannery warehouse. In back of that, on a low-lying flat swampland, wooden planked walkways led to a series of plain wooden homes and bunkhouses—quarters, I was told, for the fishermen and cannery workers and foremen. Some looked to be built over a swamp, while other cabins dotted the grassy waterfront only a few yards behind the high-water mark of the rocky berm slope. And up and down that western-most shore, directly in front of those austere cabins, a scattering of wooden gillnet boats swung easily at anchor.

In back of it all, at the head of the valley, the summer-green hills of Unimak Island rose steeply toward the volcanic glacier-studded peaks of 6,140-foot Roundtop Mountain.

We had hardly passed the village when a hundred-foot crab boat struck bottom ahead of us and began veering this way and that in an effort to locate the channel. With her propellers grinding through the silt, a brown trail of muddy water fifty feet wide boiled out behind her. Finally free of the mud and back in the main channel, the skipper began chattering over the CB set.

"Good God! I'd be damned careful about passing through here!" he warned. "Damn! I was right where those tin cans they call buoys say we're supposed to be. And we still hit bottom. What's a guy gotta do to find the channel? We hit awful hard. Hope my hull is all right. I'll bet that dulled the hell out of my props. God, I hate this stretch of water! We'll never be taking this route again. I'm getting too damned old to be taking these kinds of risks. No sir, it's the long way around next time for this old boy!"

With our crab tanks pumped dry and our hull sitting high in the water, ours was a different journey, and Lars pushed ahead through those same waters without flinching. I used field glasses to do the spotting, and managed to identify each of the buoys long before we got to them. We maneuvered without incident through the entire length of the poorly marked channel.

Lars timed our arrival at the far end of the passage to coincide perfectly with the peak of the tide, and we motored confidently across the shallow waters of the bar at the far end of False Pass. There, at the mouth of the inlet, we passed the beached and rotting remnants of a vessel that had not been so lucky. Her wooden hull beams were jutting up from the nearby sands, arching skyward like the picked-over rib bones of an abandoned carcass. Then our ship moved ahead into the broad gray plate of water known as the Bering Sea.

Some 1,600 miles wide and 760 miles tall, the Bering Sea is almost 900,000 endless square miles of water. She is bordered by the Aleutian Islands to the south, the Alaska mainland to the north and east, and the Soviet Union to the west. Surrounded by primitive shores and crowned with a monotonous gray vault of sky, it is a cold wasteland of wind and sea. As we set our course for the crab grounds several hundred miles away, I recalled one Kodiak crab fisherman's succinct description of what he had seen there: "miles and miles of miles and miles."

The Bering Sea is one of the planet's moodiest, with weather ranging from unendurable monotony to monstrous and incredible bouts of volatility. For days, this sea can lie calm and still as a graveyard, while the sky above her vast cold surface can remain cloudy for months. For weeks on end, however, violent sixty-knot winds can blow unchecked through her arena. Gusts have been clocked in excess of 130 miles per hour. Racing down out of the Siberian Arctic, storm fronts roll unhindered across two thousand miles of open tundra, polar ice, and featureless sea, occasionally producing storm waves so high that, on radar screens, crab-boat skippers have reported difficulty distinguishing waves from vessels.

While the Bering Sea can whip up some of the worst weather and waves of any body of water on earth, it is also the breeding ground for the largest population of king crab in Alaska or anywhere else. In fact, cut off from the North Pacific by the crescent-shaped sicklelike sweep of the Aleutian Islands, the bitterly cold and relatively shallow, nutrient-rich waters of the Bering Sea contain the single most productive fishing grounds (and the most dangerous) anywhere on earth. It owes its incomparable richness to the relatively shallow waters of a continental shelf that extends eastward from the coast of Alaska for more than four hundred miles and at depths rarely greater than one hundred fathoms (six hundred feet).

At its westernmost edge, this wild and productive continental shelf falls abruptly off into the Aleutian Basin, a relative pit of water. (An apt comparison would be to step from shallow ankle-deep water into the depths of a dark pool some twenty-five feet deep.)

At this precipitous edge, ocean waters several miles in depth swirl up and over the edge of the continental shelf, carrying salts and minerals and the accumulated remains of dead and decaying phytoplankton and aquatic plant life. As these currents deposit their rich cargo of nutrients, the world's largest schools of bottom fish gather. Scattered out across the rich undersea waters of the continental shelf, great cyclic populations of king crab, tanner crab, and opelio crab graze and proliferate.

The object of our fishing was the king crab, an odd and yet remarkable spiny-legged creature. For centuries prior to the advent of commercial harvesting, the king crab had lived out their harmless lives without interference from man. That changed with the coming of the Russian fur traders in the early 1700s. There is some evidence that the early native populations in Alaska enjoyed the tasty white meat of this spiderlike crustacean.

One Russian diary (only recently translated) written in Kodiak in the late 1700s claimed that "In addition to bears, there are many fox, arctic fox, halibut and other fish, as well as enormous crabs. The natives eat only the legs of these crabs, and each is as long as a man's arm. Yesterday at supper, five of us could not finish the legs of even one crab!"

According to "Mr. King Crab," Guy Powell, a lifelong Alaska Department of Fish and Game biologist, the Kodiak king crab occasionally grow to a width of six feet and weigh upward of twenty-five pounds, though the Bering Sea (red) king crab are generally smaller. Odd, ancient, and dangerous-looking, these spiderlike creatures protect themselves in adulthood with two powerful pinchers (the right pincher is as large as a man's fist) and a hard red shell covered with pointed spines. Rising up like stalagmites, these spines are sharp enough to puncture the side of a soda can.

On the bottom, the king crab locate our perforated jars full of chopped-up herring bait by means of chemoreceptors sensitive and accurate enough in the relative darkness to trace the smell of freshly chopped herring much the way a hungry human might detect a sirloin steak barbecuing in a neighbor's backyard.

The instant the scent of blood or oil reaches the crab, their antennae begin to wiggle, rising and falling before them in an urgent and excited dance, and the hunt is on.

Grazing across the ocean bottom in the utter darkness in water the temperature of liquid ice, said Powell, these bottom scavengers enjoy a wide menu. They like to feed on snails, clams, scallops, and oysters, often devouring these creatures while they are still "quite tiny . . . about the size of a button."

Hundreds and sometimes thousands of feet below the surface, adult king crab walk about on the tips of their six legs (three to a side) and, weighing only ounces in their submerged state, normally slant off to the right or left at about a forty-five-degree angle as they travel. Their eyes stand out like shiny black plastic beads mounted atop short stalks. Directly behind these eyes, sits the "small white nerve center of a brain."

They are, Mr. Powell assured me, also quite fond of barnacles, sand dollars, sea cucumbers, and sea urchins, even with their protective body needles protruding outward like porcupine quills. In addition, king crab salvage nourishment by feeding on "hundreds and hundreds of different kinds of worms" that live on the ocean floor.

Eating protocol for the king crab involves holding and crushing food with their large right claw while their second, smaller pincher moves in, pinches the prey, and tears it apart. The moment they manage to tear a piece away, they place it in their mouths, where remarkably efficient rasplike plates, top and bottom, grind the meat to shreds.

Occasionally, a king crab will tear apart a sea star or a starfish, and then a feeding frenzy may occur as other crab arrive and begin fighting over the food. Then the normally docile, slow-moving creatures may cloud the water with mud and dismembered particles of food.

While equipped in adulthood like armored warriors of the deep, king crab begin their lives in docility. According to Powell, female king crabs lay their eggs through a tiny hole called the gonopore. The gonopore is completely surrounded by countless numbers of tiny hairs, and it is here that the male spreads his sperm. "And when that happens," Powell believes, "the female 'knows' because she can sense it chemically. The presence of that sperm signals her brain that it's time to 'lay your eggs,' " which she does.

Adrift then for months in the deep ocean currents, the king crab begin their life as larvae. It is not until they have molted several times that, measuring one-sixteenth of an inch across, they are even recognizable as crabs. Then they begin their life scavenging for whatever food they can find across the ocean floor.

During their first year of life, the crabs are solitary by nature. During their second and third years, however, king crab often migrate across the bottom in yard-deep piles of similarly sized crab called "pods."

King crab are said to grow and physically mature through a series of "molts," a remarkable process in which the crab sheds its entire exoskeleton. Gills, digestive system, tendons, pinchers, and even the eyes are included. The crab begins this process by forming an entirely new shell inside the old. When it is adequately developed, the crab ingests extraordinary amounts of water. Expansion results, parting the old shell from the new, and the crab draws out backward, sporting a new and larger spine-covered red coat of leg and body armor. A king crab that has just molted, I was told, was said to feel "as soft and spongy as a balloon full of water."

While molting is essential if a crab is to grow, there are no absolutes in the frequency of the molts. While a young king crab may molt twice a year, another might not molt once during the same year. Later in life, a male king crab might molt only once every two or three years, while the younger, harvest-aged males do so about once a year, resulting in an annual growth of about an inch.

A male king crab that has not molted for quite some time is called a "skip-molt." These tired-looking creatures struggle to move at all on deck. Weighed down with barnacles, algae, and occasional ferns standing several inches high on their backs, they die quickly inside our live tanks. Knowing that in their weakened state their decaying bodies will soon poison the live and healthy crabs around them, it would be our duty to toss each one back overboard.

King crab have other unique characteristics that give them incredible healing powers and cause cannery owners headaches. Through their tough bodies circulates a blood that is white, not red. Instead of hemoglobin, their blood contains hemocyanin, which coagulates at a remarkable rate, making the relatively quick repair of serious injuries possible.

Just as a lizard might lose its tail and regrow it, a king crab has little trouble dropping a mutilated leg, and can do so with only minimal bleeding, though it would take the crab some seven molts (four to six years) to regrow it. I'd seen it often while working on deck. A king crab grabbed roughly by the legs might release one or more of his legs.

It is this same coagulative quality that gave the first processors in Alaska trouble. Early canneries discovered that if they allowed too much blood from the crustacean to find its way in with the crab meat, later, when the can was opened, they would find the contents covered by what looked like a "piece of glass." The copper of the can and the blood, it was eventually discovered, were reacting chemically to one another, and the blood was solidifying into something akin to Jell-O. Cannery people called them "strudite crystals" and introduced tin cans in an attempt to solve the problem.

Like the rest of my crewmates aboard the *Williwaw Wind,* I paid little mind to the Alaska Department of Fish and Game estimates on the current king crab populations. The season was at hand and, for better or worse, we were married to the ship and committed to the season ahead. Over odds too remote to calculate, I had come by a full-share berth (a 7 percent take of whatever the boat caught) on a Bering Sea crab boat and, boom or bust, I planned to see the adventure through.

Some biologists believed a bumper crop of as many as 60 million legal-sized male king crab were wandering the sustenance-rich undersea floor of the Bering Sea. With an average male king crab weighing in at six pounds, this meant that the flat, Kansas-like topography of Alaska's continental shelf held more than a third of a billion pounds of harvestable king crab.

If one could believe the surveys, never before in modern history had so many king crab and crab boats been present simultaneously in the Bering Sea. At the same time, never before had the canneries offered fishermen more for their king crab catch. The price being offered and the availability of king crab had virtually quadrupled over the past several years.

To compound the potential for profit, newly constructed multimillion-dollar crab boats from 110 to 170 feet in length were arriving in Alaska every month. They could turn over gear in two-story seas, while many of the crab boats of old could only button down the hatches and idle into the storm. Most advantageous of all was their capacity to carry crab between the fishing grounds and the canneries. While many of the older boats commonly packed no more than 50,000 to 150,000 pounds at a time, the live-tank holds of the new vessels could pack unheard-of loads of 200,000, 300,000, even an incredible 500,000 pounds of king crab in a single load!

Everything seemed poised for a tremendous season, barring mechanical breakdown, bad weather, injury, or death. Unbeknownst to us, the Alaskan king crab industry was about to explode into an economic bonanza unparalleled in the history of Alaska.

THREE

A day later, we began preparing to dump our entire deckload of 240 crab pots overboard. As we cut through a gray and rolling sea, closing steadily on our skipper's favorite crab-fishing areas, our crew joined in packing dozens of twenty-five-pound blocks of frozen herring from the freezer near the galley out onto the edge of the deck. Hiking along through a tunnel built beneath the twenty-five-foot-high stack of crab pots burying us, we dropped each block beside the hydraulic bait chopper bolted to the wheelhouse wall. Then we stripped the plastic from several of the frozen blocks, dropped one down into the open mouth situated at the top of the machine, and slid a long wooden box underneath it. Then Bobby hit the Power On lever.

The aluminum box exploded to life. The frozen herring block clattered noisily about inside the four-foot-long rectangular hunk of metal, banging against its metal sides like pieces of rock. The noise was deafening and efforts at verbal communication ceased during the few seconds it took to grind up each block. Then an avalanche of frozen herring bits tumbled out the bait chopper's bottom chute and fell into the wooden box.

We worked together in the limited space of the tunnel, kneeling and filling empty white plastic bait jars with the ground-up herring. These jars had snaps attached to their lids and were designed to hang from the webbing inside the crab pots. They were perforated with hundreds of tiny holes to allow the smell of the herring oil and blood to flow out through the undersea currents and attract the crab.

"We will hang two bait jars in each pot," said Bobby, speaking to no one in particular. "And with two hundred and forty pots in the stack, well, we will be needing to fill five hundred bait jars before we are through. You see?" He paused and looked at me. "You will be working up on top, big feller," he said. "You and Sig will untie the pots and have them ready to go and hook them up to the crane line when the time is right. We will launch one pot every minute, from the start to the finish." I pursed my mouth at the expected pace. "We can do it. No problem," he said, looking at me. "Oh, you better believe it."

For the next five hours, we carried frozen herring blocks from the freezer, ground them up in the bait chopper, filled the plastic bait jars with ground-up herring, twisted on their red plastic lids, and hung them in groups of twenty-five. It was a pleasant fellowship of fishermen, and hour on hour we worked silently, each leaving the other with his own thoughts about the season ahead.

We barely had finished when the skipper's voice spilled from the deck speaker. "Okay, boys, this looks to be the spot. Get the first pot in the pot launcher. I will tell you when to launch it. Keep the pots coming until I tell you otherwise."

Then I felt the ship come around on a new and specific course, and Lars's voice sounded once again. "Remove the chains now, Bobby," he said to our deck boss. Working up and down and across the twenty-five-foot-high stack of pots, we hung precariously out over the icy sea below and, like monkeys, clung with one hand to any grip that offered itself as we untied the crab pots one by one.

When we completed the precarious task, Lars acknowledged it with a single word. "Okay," he said, signaling Bobby to begin launching the pots overboard. Over the next few hours, our skipper would crisscross his favorite patch of gray mud bottom with pot strings, thirty and forty pots to a string, and each set approximately a quarter of a mile apart.

Working on top of the pot stack, high over the deck, I saw Bobby and Fritz on the deck below remove the buoys and line from inside the first crab pot. They tied the pot door shut then and tossed the buoys and lines overboard. I raised up in time to see Bobby hit the hydraulic lever and watched as the pot-launching rack tilted up and the season's first crab pot slid from the tubular steel rack and sank into the foamy depths.

I could feel the ship rolling gently from side to side as we worked, and I busied myself by racing from one pot to the next, untying the multiple tie-down lines, which, in addition to the long chains, were used to secure each pot in place. Sig trailed behind me; it was his job to attach the boom-line straps dangling down from the incredibly long arm of our deck crane to each crab pot. While every chore was important, everything on deck centered around Bobby's ability to run the crane. It would be his job to pluck each crab pot from the top of the stack far overhead and place it quickly and gently into the pot launcher on the deck below.

Back in Seattle, I had worried about the lengthy bouts of excessive drinking my crewmates were given to in their off-hours. But now at sea, they proved to be a determined and knowledgeable crew of fishermen. With Bobby guiding us through the motions, we off-loaded our mountainous deckload of crab pots, working safely and efficiently, and without argument or complaint. With all the running between strings, it took only seven hours to rid ourselves of the huge deckload of pots. And when the last of our 240-pot stack had been untied and baited and launched overboard, our skipper turned the *Williwaw Wind*'s bow west toward the ancient port of Dutch Harbor, where we would pick up our second load of pots.

Three days later, we arrived back in the Bering Sea with another two hundred pots, which we hurriedly baited and dumped overboard in thirty- and forty-pot rows. We now had approximately 440 crab pots fishing for us, and we hurried back to the open sea area where we had dumped our first load to check on our luck.

With a thin cloud cover overhead and a blinding platinum sea encircling us, and with no other ships in sight, we prepared to pull our first crab pot. The tension on deck was almost palpable as we stood near the railing and waited for the first crab pot of the season to come aboard. Our crew had spent the long months of summer sweating to prepare our gear in the pot yards of Ballard, and were now committed to several more months of labor and danger in the king crab season ahead. Good fishing, we knew, could make all the work and waiting worthwhile.

Partly because of my youth, and perhaps also because of my background as a nationally ranked shot-putter in college, I was assigned the task on deck of hooking each set of buoys and hurrying them on board. The process involved tossing a four-pronged grappling hook (weighing several pounds and attached to seventy-five feet of half-inch line) out over the buoys, allowing it to sink far enough to snag the line, and pulling the line and buoys hand over hand across the tops of the waves and flinging them up over the side and onto the deck.

This morning would mark my first throw of the season. With the entire crew watching, I felt empty with apprehension as I coiled the buoy line and prepared for the toss ahead. A miss would mean our skipper would have to reverse his engines and back down sharply, or he would be forced to make an entire loop around the crab-pot buoys and maneuver in close enough for me to make another throw. I drew back and sent the four-pronged chunk of steel and thin yellow strand of trailing line in a high arc out toward the string of buoys drifting on the sheeny gray sea. I watched as the grappling hook cleared the buoys and dropped into the seawater beyond. I allowed it to sink briefly and, pulling the hook in hand over hand, managed to snag the 5/8-inch yellow strand of polyester line between the second and third buoy. Then I yanked fiercely on the line, hauled the entire triple-buoy setup aboard, tossed it onto the deck behind me, and quickly stepped back out of the way.

Without a word, Fritz flipped the line into our powerful hydraulic block and gave Bobby (stationed at the controls) a nod. Then Bobby's hand moved against the lever. The power block jumped into motion and five hundred pounds of stainless steel began to pop and growl as it winched the crab-pot line aboard. The line spewed out from the block at three hundred feet per minute, and Fritz coiled the line into a neat pile (shot) at his feet as fast as it came at him. As he worked, the rest of the crew remained silent with worry and superstition.

76

With the silver sheen and choppy seas blocking our view, no one could see the crab pot as it rose. Then we heard it thump against the *Williwaw Wind*'s hull and, a moment later, a rectangular seven-by-seven-by-three-foot cage of steel and mesh broke the surface and rose dripping from the sea. I froze at the vision. The webbing of the cage was stretched and bulging, and the steel frame of the crab pot's door was actually bending under the burgeoning weight of hundreds of king crab, both male and female.

"Good God!" yelled Sig. "Just you take a look at this!"

With Bobby still at the controls, and Fritz and I standing next to the railing on each side of the pot launcher, we guided the pot up and over the side and into our pot rack. In a rush, we removed the crane hook from the bridle, untied the steel door, swung it open, and stood aside as more than a thousand pounds of live king crab spilled onto the deck.

Nearby, the gaping hole of the live-tank chute sat upright and waiting. Hollow and constructed of welded aluminum, this waist-high chute fit snugly into an opening cut in the live tank's hatch cover. It was into this opening that all live and legal male king crab would be tossed.

We bent quickly to the task at hand, sorting out the legal males and throwing the females and undersized males back overboard. I had hardly begun when I saw Bobby pick up a male king crab and lift it to his lips. Then I watched as he kissed the crab's spiny shell and tossed it back overboard.

"He will grow big now," he said, turning to me and grinning widely. "And next season he will come back to me and he will bring ten thousand of his buddies!"

Suddenly, Lars's voice rang out over the loudspeaker.

"Bobby!" he chided. "Bobby, what is this that you do? Does this mean that you are lonely? Perhaps we can get you taken care of in town, Bobby, but for now you must wait!"

Bobby pivoted and smiled up at Lars in the wheelhouse. Then he turned back to the sorting at hand and waded into the knee-deep eight-foot-wide pile of writhing king crab before him.

I smiled and shook my head at such a superstition. I'd seen something like it before in the oil fields of Louisiana when a grease-covered oil driller poured champagne into the hole of a new well in an effort to "call in the dinosaurs."

I picked excitedly through the pile of king crab and rejoiced along with the rest of my crewmates, for the great majority of the king crab were not only male but also large and fully legal.

"That's money in the bank, you beautiful bunch of squareheads!" I yelled. "What do you think, Bobby?" I raved. "A few hundred more of these and we'll be sitting in fat city! Oh, I can see it all now. Soon, up in neon letters in downtown Ballard there'll be 'Bobby's Bar' and . . ."

"Shut up and pitch the crab," he ordered gruffly, shaking his head and adjusting his cigar.

Out of that first pot alone we "live-tanked" 191 adult male "keepers." At $1.01 per pound, this pot alone would be worth some $1,146, netting each of us on deck an individual crewman's share of $80.22. The crab pot had taken less than five minutes to winch aboard, empty, rebait, and dump back over the side again.

Though the catch per crab pot soon dropped well below one hundred, the fishing remained excellent and, except for a few short hours of sleep each night and even shorter runs between strings of gear, continued day and night without pause. I felt as though I'd been tossed into a fisherman's dream.

We weren't the only ones who'd set on the crab. From the onset, large numbers of crab-boat skippers discovered that their catches had more than doubled from the previous year.

Hauling gear—picking up crab pots, unloading the crab inside, rebaiting them, and launching them back overboard again—we worked through the second day, halting work at 2:00 A.M. only long enough to catch three hours of sleep. Then at 9:00 A.M., we paused again for a quick breakfast of scrambled eggs, sausage, hash browns, and toast. Between bites, Bobby announced that at the rate we'd been boating king crab that morning, we'd earned more than three hundred dollars per man per hour.

Back on deck, I was contentedly tossing legal male king crab into our live-tank chute when a sharp, irritable voice cut into my thoughts.

"You just don't get it, do you?" It was Bobby. "You just don't get it." Shaking his head, he bit down on the ugly stump of his protruding cigar and adjusted the greasy puddle of a hat he wore.

"How many times must I tell you? Here you are, jumping up and down each time you toss a crab. No . . . no . . . no! You look stupid!"

I couldn't help but smile.

"Oh, so you think this is funny!" He paused, and in a show of mock impatience, stooped once again to show me. "You bend over at the waist and then you stay there till all the crab is gone. Do you understand? You keep this up till all the deck is clean. Do you understand? Do not leave nothing! Everything you see but the wood of the deck goes in the tank or over the side. Good crabs . . . small crabs. The female crabs. Broken crabs. Snails. Everything!

"And you had better learn to turn [pivot] at the hips like this. If you do not, if you stand up every time you toss a crab, you will be worn out in only one single day! No one can go up and down like you have done. Why, your back will break in two! I do not care nothing for your back, but then we will be left to do all your work and that is something I do care about, you better believe it! So be like gorilla and keep bent over, and stay there till all—everything—is off the deck. Then you can stand up and look around and be lazy. Do you understand? Do you get it now?"

He couldn't wait for my reply. Instead, he hurried over to Sig, who was contentedly coiling away, and began venting yet further disgust.

"What is this mess?" he demanded to know, pointing down at the coil forming at his feet. "What do you call it?"

For nine continuous days we worked our gear, turning over some of our pots several times—and never did the incoming tide of king crab cease.

"This is the best crab fishing I have ever seen in my life," said Bobby on the

78

seventh day as we ran to another string of gear. "And you are one lucky young man!" he continued. Suddenly, his voice took on a brief tone of resentment. "You step right off the dock and you say 'I want a job! I am a full-share man!' and you are hired, and now, in just a few weeks you make all this big money. For thirty-eight years, I fish. I fish hard all my life . . . since I am twelve, and never has this happened before!"

While the fishing was incredible, and our earnings truly unimaginable, each time I looked up from my work on deck I saw only a monotonous gray sea stretching toward the thin line of a horizon.

The Bering Sea, the boring sea, I thought to myself.

Our existence on board was one of endless work carried out in a monotonous and wearying world coated in shades of gray. There was blue-gray, light gray, blackish gray, and slate gray. Not only were the seas gray but also our deck—gray as the paint on the fo'c'sle walls, gray as our lead lines, gray as the bottom mud that dripped from the steel frames of our pots as they swung aboard, gray as the sudden patches of fog that drifted through the fleet, and gray as the piercing pigment in Bobby's steely eyes.

Occasionally during that first trip, the sun would threaten to break out from behind the desolate gray of the sky overhead. Then one might behold a choppy aluminum glare running up to a blank horizon.

Once, as I bent over a pile of writhing king crab, I sensed a physical change in the world around me. I straightened up for a moment and found myself staring far out across the waters, squinting into a splendid vision. Perhaps two miles in the distance, the clouds had parted and angled columns of amber sunlight had broken through. Where they touched down, a dazzling white island of sunstruck sea reflected back diamond-chip blasts of sunlight. Against the Bering Sea backdrop of dismal gray, the heart of the reflection glowed with a white brilliance that looked almost phosphorescent, like the arc of a weld, pure and intense, like a frozen stem of lightning. And though, seconds later, the clouds merged and closed out the light, the glimpse lifted my spirits all day, like a manifestation of hope in a sea of unending gloom.

The superb fishing continued all that first week. Rarely did the fishing decline to a point where it became necessary to move a few pots. Then it was usually only a short distance to the end of an existing string of excellent fishing gear. When we did move a pot, I was assigned the job of pushing it across one of the world's broadest crab-boat decks. There, Sig would help me tie it in place. At these times, Bobby would fill in for me at tossing the hook. Occasionally, Bobby would miss, and then it was not uncommon for Lars to become vocal about it.

Once, Bobby missed at what should have been an easy distance. He hadn't realized that he was standing on the throwing line coiled at his feet, and as the flight of the jagged steel hook came to an abrupt end in midair, the chunk of iron recoiled back over his shoulder, sailing across the deck like a piece of shrapnel.

Something on that order happened to the best of us at one time or another, but this time Bobby was not able to recoil the line and get off another toss before we slid past the buoys. Lars was forced to run the *Williwaw Wind* around an entire circle in an effort to gain a fresh angle on the buoys. When we had closed again to within easy throwing distance with the buoys in plain view, Bobby was about to give the hook line another toss when Lars called down over the loud-speaker. "Buoy, Bobby!" he warned, extending the end of his name into a long *e* sound.

Chuckles broke out across the deck. Perched on his toes and about to launch the throw, Bobby froze and a comical look of disgust flashed across his face. Doing his best to hide it from the wheelhouse, he turned to us and, with his cigar clenched between his teeth, replied: "No shit, skipper! Really? Where? Where are they? I am old and can no longer see them. Please show them to me."

Then as we drew nearer to the buoys, Sig rushed over to offer a bit of insight. Whispering into the deck boss's ear, he said, "Bobby! Buoy, Bobby!"

Fritz ambled over then and, standing beside Bobby, pointed to the approaching buoy setup, now only yards away. "Buoy, Bobby!" he warned. Then, of course, I, too, felt compelled to join in on the lighthearted harassment.

When everybody on deck had shared the news of the approaching buoys, Bobby leaned forward and, with his throwing hook cocked at the ready, exclaimed, "By golly, you boys is right! Those buoys, I think I see them now, yes sir, by golly!" And he pointed to them as if excited at having finally spotted them.

Like an unending dream in which everything goes your way, the fishing seemed almost too good to be true. For days, our crew remained suspicious of our luck. However, when, on the last day of that first trip, the skipper told us we had almost 250,000 pounds of king crab in our holds, and still the fishing showed no sign of letting up, Bobby and the rest of the crew from his homeland began to cheer up.

Suddenly, a Norwegian *shada* began blasting out over the loudspeaker, and when those from the homeland heard it, they could no longer contain their pleasure. As if on cue, they locked elbows and, arm in arm, began to dance. Round and round they spun, grown men absolutely ecstatic over their good fortune! Their knees pumped high and their feet hardly touched the deck as they danced. Then, in perfect cadence, they reversed themselves, laughing and calling out to one another in their Norwegian tongue like children at play.

Bobby had jumped smack into the middle of the commotion. Overwhelmed by the spirit of the moment, he moved about like an exuberant youth of twenty and, grinning widely, hummed impishly along with the song, all the while looking like a delighted leprechaun.

Then, suddenly, the music broke off and a voice came over the loudspeaker. "Buoy, Bobby."

FOUR

Soon after our spontaneous dance, we cleared the *Williwaw Wind*'s deck of crab pots, buttoned down our hatches, secured our power block, and made for Port Moller over on the Alaska Peninsula.

In 1977, a fleet of 121 crab boats had fished the Bering Sea, dotting the hundreds of miles of king crab grounds with an estimated thirty thousand crab pots. Now, only a year later, the size of the fleet had expanded by nearly 50 percent to 179 boats. Even more worrisome, the average carrying capacity of the new ships arriving on the scene was much greater than in 1977. Ships like the 160-foot *Pacific Apollo* could carry an unheard-of eight tiers (a stack approximately thirty feet high) of crab pots on a back deck big enough to hold full-court basketball games when in port. She could carry five hundred pots in a single load—ten times the amount many smaller vessels could pack. Such capacities had been instrumental in pushing the estimated number of crab pots fishing in the Bering Sea past 55,000, for a single season increase of over 80 percent.

On the 110-mile, twelve-hour journey into Port Moller, even those of us who refused to believe such reports could not deny our eyes. For as we ran, we were constantly forced to weave our way through other boats' gear, and came upon column after column of crab-pot strings, with their brilliant fluorescent-orange buoys stretching off toward the lonely gray horizon in either direction.

With several crabbers working in the area at the time, we passed nervously through great islands of dense fog, and felt a great sense of relief when we reached the broad entrance to Port Moller. We planned to motor upstream to one of several floating processors anchored in the headwaters of the bay, tie up alongside, and off-load what we estimated to be 305,000 pounds, perhaps the largest single boatload of king crab ever delivered in Alaska.

As we entered the bay, I climbed up into the wheelhouse and scanned the countryside with the captain's field glasses. Off our starboard, a broad and grassy peninsula of sand extended for several miles down to the sea. Off our port side, I could make out the long silver roofs of cannery buildings, a bleak scattering of gray and weathered homes, and the tall bleached masts of salmon boats stored for the winter.

Behind the village rose steep thousand-foot hills, and standing bolt upright on that otherwise-trim skyline rose several massive steel DEW-line (National Distant Early Warning System) radar towers, the broad sweep of their rusty silver-gray screens looking out over the Bering Sea like giant sentinels. In the distance stood the perennially snowcapped peaks of the Alaska Range.

Most startling of all was the sudden transformation of the land. Little more than two weeks had elapsed since we skirted the Hawaiian-like shores of Kodiak Island and the steep, rolling hills of the Alaska mainland. But now the lush green valleys and flowered highlands had been transformed into a lonely brown landscape the shade of tanned buckskin. It would remain that way for the next nine months, and except for the quick-melting blankets of snow, the bone-chilling months of arctic winds, and the short, spirit-snuffing hours of winter light, little in the landscape would change.

As we made our way slowly into the head of Herendeen Bay, we were greeted by an unsettling sight. The giant 465-foot processing ship M/V *All Alaskan* was surrounded by nearly a dozen crab boats, all in line ahead of us, and all waiting to be off-loaded.

Aboard the *All Alaskan,* processing crews were working around the clock to kill, cook, and freeze the endless deliveries of king crab. We could do nothing but wait our turn. For four excruciating days, we waited. Periodically, our skipper went aboard the *All Alaskan* processing ship and ranted and raved like one possessed. His arguments were absolutely valid. Each day, as we sat and our pumps flushed the bay water with its high content of fresh stream water over our crab, more and more of them would die—possibly thousands. But it was equally true that every other crab-boat skipper in Herendeen Bay (and in the entire Bering Sea, for that matter) faced the same dilemma. The processors were overrun with king crab.

In spite of Lars's dramatics, the *All Alaskan* off-loaded our crab when it was our turn and no sooner. Our live delivery weighed approximately 210,000 pounds, while our dead loss totaled a disheartening and wasteful 85,000 pounds.

"Never again will we come here!" scolded Lars as we pulled away from the processor. His ruddy face and long neck was stretched out the port-side window of the wheelhouse. It was a parting shot meant for those standing aboard the *All Alaskan,* and was delivered with the tone of one who had found himself on the short end of a conspiracy. "We'll be taking the next load to Kodiak!" he added, as if swearing an oath.

I loosened our bowline and the cannery foreman aboard the *All Alaskan* slid the eye of the hawser line off the steel cleat. When Lars stuck his head back inside the wheelhouse, the foreman quietly mouthed one word. "Good," he said, wearing a look of relief.

We motored down the length of Herendeen Bay then, and as we moved out into the rough and unpredictable waters of the Bering Sea, I was struck by the fact that as terrible as our dead loss had been, my first trip aboard the *Williwaw Wind* had personally netted me some fourteen thousand dollars!

We neared our crab-pot strings just before dawn. I heard my stateroom door open, and lifting my head, I squinted sleepily into the harsh shaft of hall light playing across the room. Then I heard Lars's warning voice ring out. "Drop your cocks and grab your socks, boys!" he said, snapping on the light and slamming the door.

Tossing my bedclothes back, I leapt to my feet, stepped into a faded pair of black Cisco work pants, pulled the red suspender straps over my shoulders, and stepped out on deck to relieve myself. The first pale blue light of dawn had unveiled a world strange and alien to the Bering Sea I'd known. The air hung breathless over our ship and crew, and the water languished like a glassy, un-rippled pond of oil, without so much as a zephyr of wind to ruffle it. As the sun began to show herself, the sea seemed to stir to life, and in the early dawn light, the surrounding water began to flex and shift in broad, sheeny black plates the color of blueing on a new gun barrel.

To our delight, we found that while we were off-loading in Port Moller, our crab pots had once again been overrun with king crab. And we set out imme-diately to reap what my seasoned crewmates assured me was a once-in-a-lifetime harvest.

After those excruciating days of waiting to off-load, it felt good to be turning over our gear again, laboring far from shore, or even the sight of it, with a flat, blank line of lonely horizon surrounding us. By noon, the rising sun warmed our world to perhaps forty-five degrees Fahrenheit and had changed the dark and moody hue of the early-dawn sea into a spirited reflection of the pure Alaskan sky overhead, one that was rich and exotic.

Later, ground swells began to move through the area, and as our ship moved from pot to pot, we could look up from our work and see their smooth, rolling progression. Passing through the flat mercuric blue water, they moved off toward the limitless horizon like wrinkles through a settling blanket.

That afternoon, a misty island of boiling fog appeared in the distance. As wide as the horizon, it rolled steadily toward us, and our crab-pot buoys vanished in its path. There was no way to avoid it, nor was there any way to tell how deep it lay. During my two-hour watch the night before, I'd overheard other skippers complaining about the low-lying fog banks drifting through their gear, but there was nothing to be done. We did our best to ignore it, and returning to the labors at hand, we tossed the grappling hook, sorted crab, coiled line, and chopped bait. No one on deck spoke until the fog was almost upon us.

"This will slow us some," said Bobby, rising from his work. "We will need to keep a sharp lookout for the buoys now; you better believe it."

Then the great wall of gray-white fog rolled in over us, swallowing us whole. Its face-wetting mist boiled silently across our deck at an angle directly perpendicular to our course, and as it did, the illusion it created was one of our ship moving sideways at an impossibly fast clip.

The dense fog swirled past seemingly without end, and when I looked up, the fo'c'sle at the opposite end of the deck had disappeared and the wheelhouse directly above me was fading fast. Unable to locate the buoy setup drifting above each crab pot, we drifted and waited; inside, the skipper scanned our foggy radar screen for traffic. But our luck held. In less than an hour, the smothering veil passed on and our profitable work resumed unchecked.

Other crab boats, however, had far more lengthy and serious encounters in the fog.

The skipper and crew of the 121-foot crab boat *Rondys* had been enjoying excellent crab fishing not far from our location when, that same afternoon, they decided to make the short run over to their last string of gear. It would be their final stop before heading into Port Moller to unload. En route, they ran into a thick bank of fog. Terry Sampson, a superb young deckhand from Vashon Island, Washington, was at the helm at the time.

He would later remember the *Rondys* sliding across an ocean that lay as flat and quiet as a country pond. "It felt peaceful in the fog," he said. From a distance of several miles, another vessel was closing on him from off his starboard side. Then, in quick succession, the *Rondys* sank into a blinding wall of fog and he lost the other ship on the radar screen. For no particular reason, Sampson hesitated to awaken his skipper. He would see the *Rondys* through the present predicament and no one would be the wiser.

"The skipper had always been adamant that on watch, everything be done right," Sampson told me later. "The second I lost that ship on the radar, I should have gotten him up. He was only grabbing an hour, his first in more than a day."

Sampson found it was "blind going," moving through the thick fog. But there were brief open patches and the *Rondys* passed in and out of them as it powered ahead. Then he slid into a solid wall of fog that seemed to stretch on without end, and for long minutes the sky and sea vanished altogether. He knew the other crab boat had to be close. "It was a nightmare situation," he said, and when the *Rondys* once again broke out into the clear, Sampson felt a deep sense of relief.

The *Rondys* was moving ahead across the clearing at only half speed when the sharp, steep nose of the *Paul Lin* materialized out of the wall of fog on Sampson's right. The vision still haunts Sampson: the ghostly steel ship lumbering out of the feathery gray-white fog at what seemed full speed, and the terrifying inevitability of the collision.

It was obvious to Sampson that he was going to "strike her directly amidships." Considering the superior size of the *Rondys,* and the combined velocities of the two closing vessels, he was sure that the *Rondys* would "crunch right over" the much smaller *Paul Lin.* If that happened, there was a good chance of burying the ship and its crew right then and there.

Sampson jumped to his feet and yelled aloud, "No! *Nooo! Nooooo! Nooooo!"* The intensity of his pleading denial increased as the two vessels closed on one another. Instinctively, he grabbed the port engine throttle, slammed it into emergency reverse, and opened it up, full throttle! Frantically, with adrenaline-induced speed, he cranked the rudder over hard to port and gunned the starboard throttle all ahead full while the port-side engine roared in reverse.

But there wasn't time, and when he saw that the *Paul Lin* seemed to be neither slowing down nor taking evasive action, Sampson said he could see the situation was hopeless.

The blunt steel nose of the 121-foot *Rondys* struck the *Paul Lin* on the port side of the bow. The hull shivered and both ships spun sideways in opposite directions. The feeling aboard the *Rondys* was that of an overpowering force moving smoothly through a substantially lighter one. The impact was so slight that few others than Sampson and the skipper felt it.

When crewman Dave Capri reached the wheelhouse, both ships were floating stern to stern. His own skipper, Vern Hall, was already there.

"Where are they?" asked Capri.

"They're back there!" said Sampson.

Capri looked over at the *Paul Lin.* "It looked like someone had taken a giant can opener and just pried her open," he recalled later. The torn hull metal formed a gash on her port side beginning two feet above the flat waterline and ran up to her bow. He could see right in.

While the blow to the *Rondys* was slight, for those on board the *Paul Lin,* the impact was explosive. Jim Taylor, a crewman on board the *Paul Lin* at the time, recalled the "screeching sound of metal scraping and crunching . . ." and claimed it "sounded like we had hit a rock or run aground. Instantly, everybody was up and running around looking for their survival suits!"

There were two sets of bunk beds in the forward stateroom where the bow of the *Rondys* broke through the port-side wall. Two men were resting in the bunks mounted against the wall, the skipper of the *Paul Lin* told me later. "One was thrown completely across the room and ended up in bed with a fellow crewmate who had been asleep in the bunk straight across from him. The other, the one in the top bunk—next to the bulkhead—nearly duplicated the feat, but bounced off the side of the upper bunk . . . and ended up on the floor."

Another crewman, who'd been busy making a Crab Louis salad down in the *Paul Lin's* galley when the collision occurred, ended up in the dishwasher.

Later, the twenty-three-year-old skipper of the *Paul Lin* described for me what had happened. "I was looking down into the port radar at the time," he said, "and their anchor, which was right on top of their bow, hit me in the head and that's what knocked me out! Their anchor came through my port window and hit me right in the head. And then their bow came up and creased our port side. [The gaping hole] was probably about a fathom wide at the top and tapered down to just below the waterline."

Luckily, it was a flat-calm day, one of those rare Bering Sea days when nearly all wave action was absent, with little or no swell running. Using some crab-pot doors and some mattresses taken from the men's bunks, the quick-thinking crew of the *Paul Lin* patched the lower portion of the hole nearest the water.

Sampson knew that, for the *Rondys* skipper to have allowed a crewman to stand a watch by himself, meant that he trusted his judgment and abilities. "When the accident happened, I felt I'd really let down a lot of people," he recalls.

After conducting a thorough search of the ship for damage, the crew of the *Rondys* gathered in the galley with their skipper, Vern Hall.

"We've got a serious matter on our hands," he began. "I understand that everybody's okay on the other boat. Nobody was hurt; that's the main thing. It's not the end of the world. But there's going to be a whole pile of red tape arising over this for a long time.

"So what I want each of you to do is go back to your room and think about this. Think about things, and try to remember things the way they were. And I don't want to ever detect any lying, nor any collusion. Just tell exactly what you guys saw. And remember, the only firing that'll go on over this will be because you just don't tell everything. I don't want any exaggerating, and no lying. We'll stand by the *Paul Lin,* then get patched up and get back to fishing as soon as possible."

With the *Rondys* leading the way, the *Paul Lin* hobbled across fifty miles of water and into Port Moller for emergency repairs.

FIVE

Once, as we turned over our gear that second trip, we came upon unmistakable evidence that someone had tampered with our gear and lifted our crab pots ahead of us. We could tell by the way the bait jars hung, the way the door ties were wrapped, and by the undeniable fact that the crab pots were empty, while the pots on either side of them netted between fifty and seventy male "keepers" apiece.

Crab thieves: I could just imagine them skulking through the night in their phantom ships, scavenging off one crab boat's gear and then another for seventy or eighty pots' worth of hot fishing. Five hours was all they needed. Sneaking about under the cloak of darkness, shrouded by a concealing veil of fog, they would keep a sharp eye glued to the radar screen as they lifted the crab from the pots of honest fishermen. They would flee at the first sign of an approaching ship, these men without consciences, and make off with as much as fifty thousand dollars' worth of king crab in a single night's "work."

"What kind of a man would run another man's gear and take his crab?" asked Lars that afternoon over the loudspeaker. Those of us on deck had no answer.

The fishing was sensational that second trip, and we were well on our way to boating sixty thousand pounds of king crab in a single day when I made a life-threatening mistake. The weather had turned sloppy once more. I had just rebaited the pot in our rack, tied my side of the door closed, and bent over to begin sorting through the writhing pile of riches at my feet.

"Okay," sounded Lars's voice over the deck speaker as he shifted down and accelerated ahead.

87

Bobby had been standing near the block with his hand on the hydraulic controls. At Lars's signal, he launched the pot over the side, tossing the first coil of line over with it. Then a snag appeared in the first few fathoms of the second coil of line lying on the deck. Bobby moved to the railing and began to untangle the line and feed it overboard as fast as his hands could fly.

I straightened up with three king crab in each hand and was about to launch them into the tank chute when, without realizing it, I amateurishly lifted my right heel an inch or so off the deck. At that instant, a small wave broke over the side, washing the tentacles of the tangled line across the deck and around me. The loop closed on my right ankle like a lasso and, in a flash, I was jerked from my feet.

Lying on my back, I fumbled with the line but was unable to free myself. On the opposite end of the line, a 750-pound crab pot was free-falling down through the depths. With the slack in the first shot of line about played out, the line looped around my ankle began to drag me with a steady acceleration across the deck.

Then I heard a warning scream: "Whooooo!" Instantly, the growling diesels fell silent.

"Man in the line!" hollered Bobby.

Then I heard the emergency clank and grind of gears as Lars forced the ship into reverse. The deck shook violently and the water boiled out from under the deck with a foamy roar as he gunned the twin engines full out. I could feel the huge ship hesitate as her ponderous weight and forward momentum battled with the furious reverse thrust of her twin propellers.

Compromising himself completely, Bobby leapt into the potentially deadly bite of the line, positioning himself between the railing and the pot line racing over the side. To my astonishment, he intended to single-handedly slow the overboard play of the line, allowing me time to gather myself and free my ankle. But when he grabbed the swift-moving line, his neoprene gloves gripped suddenly and his arms were nearly jerked from their sockets.

With certain death awaiting me over the railing, I tried to stand and clear my foot. Then Bobby was beside me. Gripping my tangled ankle with one hand, he pushed against my chest with the other and forced me back down onto the deck.

"Sit down! Sit down!" he insisted, the emphatic tone of his voice mounting.

Bobby had managed to slide the ever-tightening line from my ankle down to my foot when the line gripped suddenly, pinching my foot with unbelievable force. Bobby's face looked distorted with worry. And as I lay helplessly on my back, his hands tore at the snarl of line and clutched at my boot with a frantic intensity.

Then, like a squeezed melon seed, my boot shot skyward and flopped back to the deck next to the railing, its rubber mutilated, its steel toe twisted by the pressure.

Bobby said nothing. Shaking his head, he went back to work. Sig began readying the throwing line, preparing for the next throw, while Fritz Boettner guided the rest of the line and the crab buoys over the side. No one, not even the skipper as he gunned the engines and moved toward the next set of buoys, said so much as a word.

Giddy with excitement, I started to laugh.

"Oh, you enjoyed this, did you?" queried Bobby.

"No, not one bit," I assured him. "But I'd given myself up for dead. It just feels good to be alive again."

SIX

Day and night throughout that entire trip, our crab pots rose to the surface bulging with king crab, and I experienced much the same greed and joy that a lucky Klondike gold miner must have felt when the wash of his gold pan finally showed color and he lifted his first nugget.

Even as our live tanks filled, we knew the real problem would be to find a cannery we could deliver to before the crab died in our holds. After the painfully slow off-loading of our first (and last) delivery of king crab to the floating processor in Port Moller, we were faced with the perplexing problem of locating a new market. In the Bering Sea during the overly abundant season of 1978, this was a monumental task.

The sheer unexpected volume of crab had flooded local canneries. Dozens of boats, their live tanks packed with 100,000-pound loads of king crab, had descended within hours of one another upon the processing facilities in Dutch Harbor, Akutan, and Herendeen Bay and were threatening to bury them.

In Dutch Harbor, boats were anchored off in Unalaska Bay, strung out in lines that would take a week or more to process. And if you had no previous agreement with a local cannery there . . . well, good luck, pal. With all the waiting, losses were reportedly mounting into a serious problem even in the salt-rich waters of Dutch Harbor. The waste would run into the millions of pounds.

But then there was Kodiak. Given good weather, a modern crabber could run there, deliver, and be back fishing in the Bering Sea in about a week. It was rumored that fishermen around Kodiak Island were having a dreadful crabbing season and that the canneries there were begging for crab and offering sky-high prices. There was even talk of a price war. But if that was the case, why wasn't everybody heading for Kodiak?

The main reason was that the seven-hundred-mile journey was filled with risk, something we calculated over and over. Leaving our gear unguarded in the Bering Sea, we would have to cross over the False Pass bar at high tide, sweat through the poorly marked shortcut, weaving back and forth and scrutinizing our fathometer's every blink as we searched for the main channel. We would then have to slip delicately past hidden sandbars, and if we wandered and ran aground, we would have to float ourselves free by pumping our live tanks dry, something that would guaran-goddamn-TEE the crushing of thousands of our crab.

Once on the Gulf of Alaska side, there would be storms to contemplate, and crossing the south end of Shelikof Straits through large or jostling seas would mean even more crushed king crab. Only the week before, the crab boat *Odyssey* was said to have arrived in Kodiak with thirty thousand pounds of dead crab on board.

We were only six days into the second trip, when with approximately 200,000 pounds of fine live king crab already in our live tanks, Lars spoke to us over the loudspeaker. "Make certain the live tank bolts is tight and the deck is clean, boys, 'cause we're headed for Kodiak with this here load."

Back inside and faced with the yawning distance ahead, we discussed at length our skipper's judgment at making such a long and risky run. If the scuttlebutt about the prices offered by Kodiak's canneries was true, our present load would be worth $350,000, delivered.

Vulnerable as we were during that first journey to Kodiak, we were lucky, too. Over the next few days, we skinned our steel belly in the tricky shallows of False Pass, nearly ran down a giant buoy marker in Unga Narrows, and, powering through a nasty chop at the mouth to Chignik Bay, turned east at Wide Bay. Then with the wind at our back, we ran with a tall and accommodating sea the entire way across Shelikof Straits.

The journey was mentally and physically wearing. We worried about our crab. We worried about the weather. We worried about the gear we'd left behind unguarded in the Bering Sea, and about the leak in our exhaust system, and about the cost of fuel on so long a run. In a way, we were like Klondike miners who, having struck pay dirt, now had to contend with its safe delivery. And whenever we slept, it was for only a few fitful hours—the shallow, twitching sleep of the greedy.

Thirty miles off Cape Alitak—on the south end of Kodiak Island—we passed close by a rusty old schooner working their king crab pots, and we slowed to watch. They looked as if they were averaging about two king crab per pot. I felt for the skipper and crew, as they were only two days from the incredible king crab fishing going on over in the Bering Sea.

The contrasts we had witnessed were extreme. In the Bering Sea, fishermen were boating about 2 million pounds of crab per day, more than ten times the daily deliveries to the canneries in Kodiak.

With 200,000 pounds of live and pinching king crab on board, we no sooner turned the corner at Cape Chiniak, only a few miles outside the town of Kodiak, than the bidding began.

"Williwaw Wind! Williwaw Wind!" called out one cannery rep over the CB set. "I understand you've got quite a load of crab on board! Have I heard right?"

Lars stood and grabbed the microphone from the ceiling mount overhead.

"Williwaw Wind back to the caller. Yes, yes," he began. "We have done pretty good on this trip. I would say we must have, oh, about two hundred of the big ones on board here right now, okay?" He paused. "What is the king crab going for here in Kodiak?" asked Lars.

"B & B [the cannery] back to the *Williwaw Wind.* Well, Lars, the pickings have been pretty slim for fishermen around here this season. Right now, we're looking at about one dollar and fifty-five cents a pound." He paused. A collective sigh of relief sounded in our wheelhouse. The price was ninety-eight cents a pound back in the Bering Sea when we left. Lars was nervously licking his lips when another crab buyer broke in.

"Lars, this is your old pal Peter Boteman calling from the cannery here at the other end of the Row. And we're offering a dollar sixty a pound and a night out on the town for you and your whole crew. We'll even toss in all the bait you can use! What do you say?"

Having awaited his chance, yet another voice and offer leapt from our black CB squawk box. "Lars, we can go . . . oh, how about a dollar sixty-three a pound!" the voice said. "And we'll throw in a dinner at Solley's and a night out on the town for you and your whole damned crew! What about it now?"

In the wheelhouse, we stood silently listening. After the crowded indifference of the Bering Sea processors, we were taken aback by our sudden importance. And as Lars tethered the flood of incoming offers (at prices nearly double those being offered in the Bering Sea), smiles broke out across every face— except for Lars's. Nervously licking his lips, he remained on the radio and remained uncommitted as the canneries bid the delivery price into orbit.

Even the weather seemed to be courting us into town and, aided on our way by the smooth form of a following ground swell, we rode effortlessly across Chiniak Bay. Then we turned ninety degrees to starboard, swung around the corner buoy, and idled along Cannery Row and St. Paul Harbor.

The docks were void of activity, and with our pick of empty dock space, we idled up in front of the ancient wooden hull of the *Skookum Chief* to unload our catch. They had won the price war with a final bid of $1.68 per pound.

We had hardly secured our hawser lines when the cannery crews scrambled down the dockside ladders and onto our deck. In minutes, a high-strung crowd of "cannery rats" popped our first hatch cover. Leaping down into the hold onto the backs of our king crab, they began to pitch crab at a furious rate. Shortly, a dockside crane lifted the first brailer (a giant webbed basket, carrying approximately fifteen hundred pounds of crab) up over the side and onto the cannery dock.

Reading the scale, a cannery foreman standing on the dock got things under way, calling out, "One thousand, four hundred and seven!"

"Got it!" I called back. Scurrying inside to gather my notebook and pen, I climbed topside to join him and to check the weight of each brailer of crab as it left our ship.

Twenty-eight hours later, we were off-loaded and making our way back to the Bering Sea. Our delivery had totaled more than 195,000 pounds. At the final cannery offer of $1.68 a pound, we had grossed $327,000 in a single trip. After deducting the cost of fuel, bait, and food, my 7 percent share topped twenty thousand dollars!

As we slipped out of the harbor, we passed the small, weathered form of the *Royal Quarry* tied to the B & B Cannery docks; and I was suddenly struck by how small she looked. Her seventy-one-foot length, from bow to wheelhouse to stern, could have fitted easily onto our deck, with ten feet of walking space to spare at either end.

Was that possible? When I first stepped aboard her some eight months ago, she looked like the queen of the fleet, the material evidence of a greenhorn's prayers. Since then, I had grown increasingly proficient with the basic knots of our trade, and now could tie them in rough seas and heavy winds in only seconds. I'd learned to coil crab-pot line into precise piles as fast as most hydraulic blocks could reel it aboard, to manhandle seven-hundred-and-fifty-pound crab pots, pushing them into position or across the broad, wooden slats of our huge deck at speeds, my deck boss assured me, few deckhands could match. I could read a chart, plot our course, calculate our position, and, in the howling black darkness of a stormy night, safely interpret the converging figures on a radar screen. Our skipper could be assured that I would never fall asleep during my watch, no matter how tired I was. Now, as we moved past the *Royal Quarry,* I felt much the same involuntary pangs of melancholy and mixed loyalties a happily married man might feel, who, wife on arm, happens to pass a beautiful woman friend and former lover.

At the farthest end of the harbor, we passed by the Sea Land van dock. As always, there were freezer vans full of processed king crab meat stacked high and waiting to be craned aboard a freighter and shipped to the mainland United States; their cargoes of gourmet meat were bound eventually for the opulent parties of the wealthy eastern elite.

SEVEN

Though Bobby did his best to keep us busy, our return trip was filled with boredom and apprehension. We had involved ourselves in a long-distance venture, somewhat similar to loading up in Portland, Oregon, and proceeding south—day and night at eleven knots per hour through seven hundred miles of wilderness—to San Francisco to off-load. A lot could go wrong on such a voyage.

Aboard the *Williwaw Wind,* we did our best to avoid our grumpy skipper. During the days, we worked on deck to repair several pot frames that had been bent or dismembered in the muscular shuffle of fishing. We welded them back together and stretched black nylon webbing over their frames. We untangled and recoiled ⅝-inch lines, and strung together new buoy setups, all the while sliding past hundreds of miles of wilderness sea and shore.

One morning, we were working on deck when we passed Puale Bay on the Alaska Peninsula. Motioning toward the mouth of the bay, Sig explained that several weeks before (on the night of August 28, 1978), the shrimping vessel *Jeffrey Allen* had run into trouble. While the ship drifted quietly at anchor and its tired crew slept, something had gone terribly wrong.

Rusty Slayton, the well-known skipper and owner of the *Jeffrey Allen,* told me later that he was asleep in his wheelhouse stateroom when it began. He would never know why he got up when he did, but, rising from his bunk, he climbed the short stairway and made his way to the back door of the wheelhouse. It was then that he realized the ship was rolling over.

"Get up! We're in trouble!" he screamed to his crew, asleep in their bunks below. There wasn't time for anything more, for even as he spoke the ship rolled.

He recalls that he "never had a chance at the radio" as a wall of seawater exploded through the wheelhouse windows and washed him out the back door.

Slayton drifted alongside the overturned vessel for a time and tried to get his bearings. He could hear the auxiliary engines running and, swimming to the other side of the vessel, saw the glow of the ship's lights still burning beneath the surface.

With air trapped inside, the overturned *Jeffrey Allen* remained afloat, with one row of her portholes showing just above the waterline. As he climbed up on her hull, Slayton detected a thumping sound. Peering in through one of the lighted portholes, he spotted his thirteen-year-old son, Jeffrey, and two other crewmen. They were standing on the galley ceiling, trapped inside the sinking ship.

When the group inside spotted Slayton, they stretched their arms toward the porthole overhead and began to slap and pound on its glass. That particular porthole was designed to open inward from the inside, but it was sealed by the tremendous force of the air bubble trapped inside with them, and even the frantic strength of the pulling, pounding crewmen couldn't budge it.

Slayton understood the problem. He knew that if the seal could be broken even for an instant, the window could be opened. But how? Clad only in his underwear, he had no pocketknife, and what tools he did own were trapped inside the overturned ship along with his son, brother, and brother-in-law.

Just then, he spied a bin board floating by and snatched it from the water. Using it as a battering ram, Slayton brought one end of the board down hard against the porthole window, and through repeated efforts, he finally managed to break out the glass.

The moment the seal was broken, the small round window flipped open and air began to gush out . . . and the *Jeffrey Allen* began to sink. At the rate it was going, it would disappear below the surface of the bay in only seconds.

Slayton's brother came out first. It was a tight squeeze, passing through the tiny porthole, and in the effort he scraped himself from head to foot. Waiting on the surface, Slayton grabbed the lanky 160-pounder and helped him to his feet.

Then as the two men stood on the cold steel hull, the eighty-six-foot ship began to disappear into the depths beneath them. Slayton felt the water rising over his feet and then past his knees. Air billowing out of the sinking hull roared all around him and seawater began to pour down onto his son, Jeffrey, and his brother-in-law, Don Corzine, and he lost sight of them.

Inside the ship, with the water roaring in through the porthole opening overhead and the ship sinking like an eighty-six-foot steel coffin, the youngster hesitated. The burly Corzine wasted no time in intervening.

"I can't make it! I'm too big!" he told the youngster. "But if you're going to get out alive, you gotta go now, 'cause if you don't go right *now*, you're not going to get out at *all!*"

"There would have been no chance of finding him," Rusty Slayton recalled, had Don Corzine, a husky 230-pound man of thirty, not hoisted Slayton's thirteen-year-old son up to the porthole overhead.

The ship was going down fast, descending into ninety feet of seawater, when Slayton dove under the surface and reached through the porthole for his son.

Clad only in his underwear, Slayton was too taken with the task at hand to notice the chilling cold of the water. He dove twice without success, but on his last dive reached through the porthole, grabbed his son by the hair, and yanked him to the surface.

When Rusty Slayton and his son again appeared on the surface, the ship's fully inflated life raft popped up right alongside them. Stroking to it, the three fishermen crawled aboard. Then the *Jeffrey Allen*'s lights blinked out and the ship disappeared into the deep; the popular and heroic young Don Corzine remained trapped inside.

Those adrift in the raft could see the entire shrimp fleet resting at anchor off in the darkness. But they were too far away now and there was no way to paddle against the breeze blowing out of the bay, so they drifted and rubbed each other's bare legs. They found some flares stowed on board the raft, ignited one, and waited.

Early the next morning, in the predawn darkness, the first shrimp boat pulled anchor and moved toward the open water of the ocean past the mouth of the bay. The ship was the high-line shrimping vessel *Cape Fairwell,* owned and skippered by Bert Parker of Newport, Oregon. Bert was nearly out of the bay when he spied what he thought was a flashing light.

He had no idea what it was, he said later, so he turned his ship to port and shined his mast lights in that direction. In the distance, he caught the reflection of something orange.

"Hey!" he shouted over his CB radio to the rest of the fleet. "Does anybody else see that?"

Another skipper replied back quickly, "Yah! Yah, I see it, too!"

But the strange part, Bert Parker would recall, was that he was positive the voice that came back to him was the familiar voice of his fishing buddy, Rusty Slayton. "I was sure of it!"

Parker moved swiftly to pull the three survivors aboard the *Cape Fairwell.* With one crewman reportedly missing, the fleet mobilized quickly. Captains roused sleeping crews, who scurried to pull their anchors and join the search. Despite the best of their combined efforts, though, Don Corzine could not be found.

Later, Alaska State Trooper Lt. Bob Lockman and his patrol boat *Vigilant* arrived on the scene. Using his sonar, Lockman idled back and forth across the bay scanning the bottom for the sunken vessel. After twenty hours of searching, he located what he believed to be the *Jeffrey Allen.*

The *Vigilant* crew lowered a grappling hook over the side and managed to snag the sunken ship. Lockman clinched the line tight and then sent two divers down through ninety feet of bay water to check out their find. The *Jeffrey Allen* was sitting upright in the soft mud of the bottom. Her outrigger poles extended from her sides at forty-five-degree angles and her shrimp net was still mounted at the ready, coiled around her reel. Resting there, motionless in the moody green depths, it seemed that at any moment she might come to life and, setting her net out behind her, begin her next tow.

Inside, the divers found the body of Don Corzine beneath the dining room table. They also discovered a pocket of air in the wheelhouse. If Corzine could have found his way there, he might have been able to survive for a time. "There was air halfway down the wheelhouse windows," said Rusty Slayton.

As Puale Bay slid past us, I tried to imagine the impact of such an event on the lives of others. Over the past year, more than thirty fishermen had lost their lives at sea in Alaska, leaving behind parents and wives, children and friends like the wife and newborn baby Don Corzine left back home in Crescent City, California. Rusty Slayton left the commercial fishing industry in Alaska, vowing never to return.

EIGHT

Upon our return to the king crab grounds, the Bering Sea now seemed like a different body of water. The ship rolled and the deck leapt, and the wind and waves worked at us without pause.

With the blustery weather, the mood of our skipper took on a more demanding and irritable tone. Yet through it all, the attitude of our almost pure Norwegian crew was one of quiet submission. It was the way of life on deck. The rules had always been simple. When the "old man" barked, you humped it. And if he got a little carried away with the scoldings, ridicule, or sudden bouts of temper, you looked the other way and did your best to ignore it.

But there were times when the barking went a little too far, and then there were little rebellions. Once during the season, Lars made a big thing about the fact that a few of us used catsup on our steaks. "It is one stupid way to ruin the perfectly good meat on your plate, I think you should know!" he stated adamantly one night at supper. Then, as if on cue and without saying a word, the entire crew drowned their steaks in a cheap bargain-brand catsup. Lars stormed out of the room, infuriated.

Another time, all day as we worked, Lars refused to turn down the volume of the loudspeaker that echoed his many bits of advice and criticisms out across the deck. That night, someone cut the speaker wire feeding down from the wheelhouse. The next morning as I tied the doors to a baited pot and prepared to launch it, Lars's head poked out the wheelhouse's side window and I heard him yell, "Can you fellas hear me okay!? Is the speaker working? Can you hear me?"

He drew his head back inside then, and I could see him mouth the same words over the severed speaker system. Then he poked his head outside and yelled, "Can you boys hear me, *now?*" With a smile and a wave, we assured him that we could. "Certainly, Lars. Of course, Lars. Oh, absolutely, Lars."

"Oh, yah!" hollered up Bobby. "This is just right!"

Happily shaking our heads in affirmation, we went back to the heavenly silence of our work. Stealing a glance, I could see Lars through the wheelhouse windows. He was gripping the microphone in one hand and trying to adjust the system's dials on the ceiling with the other. Each time he spoke and we did not answer, you could see his arms flailing in wild gestures and his head shake in vein-popping spasms of frustration.

Minutes later, we looked up to see Lars carefully tracing the speaker wire, following it from the wheelhouse all the way to the deck speaker. It didn't take him long to find the source of the problem and splice it together, though the volume of his communications noticeably diminished after that.

Suddenly, Bobby turned to me. "Did you do this thing?" he asked, privately pointing to the cut wire.

I hesitated. "Well now, Bobby, I do sleepwalk from time to time. I'm not admitting to anything, you understand. A man could get fired for such a thing, but there's just no way of knowing what a guy's likely to do at such times as that."

When Bobby turned away again, he was grinning widely through a bite of cigar.

One night, while working in rough seas, halfway through that third voyage, we were launching crab pots overboard when Bobby collided with one. It struck him on his right side and cracked a few of his ribs, though for a time he flatly denied the injury.

I was moved by his personal struggle to keep up and carry his share of the work load. Though he did his best to conceal it, the pain was obviously severe. Drawn and ashen gray in the face, Bobby looked decades older than his age. But he was a clever man and thereafter worked with his back to the wheelhouse, though I often saw him flinch and grimace in spite of himself.

Still, Bobby's rib injury did significantly impair his movements. The inhalation of deep lungfuls of air was impossible for him. While most men would have retreated to their bunks, over the next five days Bobby remained on deck, his face drawn and his cheeks hollow. With each effort, he would groan and fight to subdue the flinch, his face full of pain and hopelessness. But he was a good man. Even in his present limited capacity, we knew our labors on deck would increase dramatically without his efforts and guidance, and we felt lucky to have him.

Often, during short runs between pot strings, we would stand in the shadows in the lee of the wheelhouse and huddle out of the wind. On one such run, Bobby pulled me aside. The thin layer of flesh stretched across his small ashen face looked almost embalmed.

Out of sight of the "old man" in the wheelhouse above, he was holding his side. His breathing came in short puffs. He labored to talk.

"Spike, the skipper must not know of this!" he stated secretly, whispering.

99

"If you quit or get fired," he continued, "you can go and get work logging again, or maybe fish on another boat. Oh, you are a young bull, don't you see, and you can work here, or you can work there. But I . . . I am small. I am fifty-four—an old man out here!" He squinted through the pain as he gathered another lungful of air.

"Please, do not mention this thing to Lars! I don't know nothing else. I get fired, and where do I go? I have nowhere else to go to. Do you see? I can't do nothing else!"

He had managed to keep his dignity as he pleaded, but somehow I still found myself looking down and away. I shook my head.

"You've got no worry coming from me, Bobby," I replied. "Hell, that's no problem."

Bobby seemed relieved, nodded once, and leaned back against the wheel-house wall. Then I saw the tip of his cigar glow and heard him catch his breath. Two days passed before we spoke again.

All evening, the winds had been kicking up, and by nightfall the seas had turned rough and unpredictable. As if the miserable wetness wasn't enough, islands of fog began moving through the area. In the rough weather, Lars began having difficulty locating our buoy setups; and as he searched, cutting this way and that in the deepening troughs, we backed up and hid from the wind and waves, huddling once again in the lee of the cold flat steel face of the wheel-house.

Perhaps appreciative of my silence and support, Bobby and I talked again that night. He told me of the time he had outfaced a bully on deck, a man more than twice his size.

"He would not leave me alone, you see. So I took my knife, and I turn on him like this"—he raised his right hand as if to drive home a dagger—"and I say to him, 'You want it? Well, do you? Because I promise you this, I will give it to you! You better believe it!' "

His story suddenly stopped. "Well, did he back off and leave you alone?" I asked.

Bobby nodded. "He never would have nothing to do with me ever again! He thought I was crazy."

"Would you have stabbed the guy?" I pried.

"Oh, you better believe it!" he shot back, his voice thick with the emotion of reliving the event.

Then Bobby told me of the time when, more than fifteen years before, he'd been struck with a crab pot. Seriously injured and lying in his hospital bed, he recalled the day the skipper came to visit him. Without a word of concern as to his condition or the cost of his medical bills, the man had gathered up Bobby's rain gear, saying, "Well, you won't have no use for these oilskins for a while, now, will you?"

He described a man he once knew who coiled bare-handed all his life. "Oh, he was tough, like iron bark," recalled Bobby, pointing down to our deck. "And you could not tell him how to do it no other way, you see. But when he got old, his hands was bent like bear claws, and hurt him bad and were no good for nothing."

Then Lars announced that he had spotted another set of buoys. I hurried over and began coiling the grappling-hook line in preparation for the throw. Bobby followed. He appeared to be lost in thought. Suddenly, he turned to me and spoke sharply. Removing his cigar from between the pinch of his teeth, he said, "How you get a name like Pike?" That was the way he pronounced Spike. His head twisted to the side in wonder. "No one should have a name like Pike! I knew a bulldog that once had such a name!"

When, smiling, I made no effort to explain, he turned, shook his head again, and stomped away to face matters far more worthy of a deck boss.

During that same trip, a freak wave surprised both our skipper and crew. We'd been quartering into a high, loping sea all day, when, just after nightfall, Lars's warning voice came distinctly over the deck speaker: "Wave" was all he said.

Though perhaps only eight feet high, to the eye of one caught directly in the path of it, the wave looked fifteen. The "poop-sweeper" seemed to come out of nowhere and slapped us down and washed us across the deck like the right hand of Jehovah. The power and brutality of the impact and the sudden feeling of helplessness were unlike anything I had ever known.

We were bent over sorting crab at the time. I saw Bobby acknowledge the wave by lowering and tucking his head. He braced himself and froze rigidly in place. There was a moment of panic and confusion as the leaden blanket of icy water broke down over us. The impact knocked us flat and sent us sprawling. As we tumbled inside the foamy body of the wave, the bitter cold of the water seemed to reach every dry portion of clothing and skin simultaneously, flooding through my rain gear with a jolting rush that left me gasping for air.

As for air, there was none, and as the water began to ebb, I found myself sprawled on my back and struggling to rise against the undeniable weight of something on my chest. I looked up through a depth of seawater and saw Bobby perched sidesaddle across my chest. He clutched the front of my rain-gear jacket and I could feel his hands jerking this way and that as he fought to maintain his balance and remain above the current rushing over the deck— and me.

I tried to scream, but when I opened my mouth to protest, I gagged on a violent rush of water and the word bubbled out in a high-pitched, choking sputter. Furious at being caught in such a predicament, I propped myself up on my elbows, lifted my head clear of the rushing ice water, and yelled, "Get the hell off me, you square-headed bastard! You're drowning me!"

Embarrassed at his chosen tactic for survival in such a situation, Bobby scampered to his feet without a word and rushed inside, clutching his side in pain. After hanging my dripping gear up to dry, I soon followed, treating my shivering body to the pleasure of a dry change of underclothes.

We shut down that night at two in the morning, and were awakened for breakfast at 5:00 A.M. It was then that the captain spoke to us.

101

"Now you boys took some water in over you last night. You've got to keep a better lookout." It was all he said.

After wolfing down a meal of pancakes and eggs we donned our rain gear. We had hardly reached the open deck when Bobby retorted, "Why should we have to be the ones to keep a lookout all the time? Who is driving this here ship, anyhow?"

The next day, I was bent over and sorting through a pot pile of legal and sublegal crab. I had a natural working rhythm. Pivoting at the hip, I tossed each crab either over the side or into our live-tank chute and, as Bobby had instructed me, dealt with each crab as I came to it. I was launching a fairly steady stream of garbage crab (undersized males and females, many with giant rust-colored clusters of eggs clinging to their undersides) over the side when, apparently without thinking, Sig stepped directly into my line of fire.

He had just returned from lunch and had approached me from behind, coming upon me without a sound. A timely cough, sound, or hand on the back would have let me know he had arrived, but he was caught up in another thought at the time and had given no warning. He bent over between me and the railing just as I launched a slightly undersized pair of male king crabs toward the high side of the railing.

The creatures weighed four or five pounds apiece, and their sharp and spiny red backs slapped squarely against the left side of Sig's face. Work came to a quick halt as Sig yelped, shook his head, and jumped back. As he stood, thin trickles of blood began to flow from each puncture.

"Damn it, Sig! I didn't even see you coming! God, I'm sorry about that!" I apologized. "What can I do?"

Sig knelt beside the live-tank chute and splashed cold salt water across his face. Then he jumped back and stood shaking his head. "Forget it," he said, wincing at the salty wash; and, mad at himself, he went back to work without a further word.

Minutes later, Bobby saw me studying Sig. Blood was still draining from a few of his puncture wounds. Bobby edged close to me and whispered three words: "Not your fault."

At dinner that night, Lars noticed the fresh pattern of scabs across the left side of Sig's face and asked him, "What happened to you?"

"I stepped where I should not have" was all Sig said.

Our crew was not the only one to suffer various misadventures during that boom days king crab season of 1978. Tragedy and injury occurred on an almost daily basis and neither veterans nor greenhorns were spared. One crewman suffered a broken leg and his crewmate sustained a head injury from a falling crab pot. Another deckhand was crushed between the ship's bulwarks and the pot launcher. One crewman close to us lost several fingers when they became tangled between the line and the power block. Another had his hand crushed between the railing and a swinging crab pot. Still another was thrown across the deck with enough force to dislocate a hip when he tried to slow 750 pounds of runaway crab pot.

One crewman suffered a two-inch cut on the back of his head when he was hit by a crane ball. He lost consciousness and went into shock and had to be hoisted off the ship's back deck by a medi-vac (medical evacuation) helicopter, which flew him to Cold Bay, where a waiting plane rushed him into Anchorage.

While such injuries required the Coast Guard's immediate attention, there were countless accidents serious enough to justify time off from most any civilized job back on land but that at sea went largely ignored, if not unattended to.

A failing body wasn't something about which a fisherman bragged. Physical breakdown was a betrayal of one's body and, like the loggers and fishermen I'd known in southeast Alaska and the Pacific Northwest, one felt at once vulnerable and ashamed. Swollen joints, cuts, contusions, sprained ankles, water on the elbow, wrenched backs, and aching knees—injuries often too serious to permit sound sleep—went the way of Bobby's cracked ribs; they were ignored or denied.

I had always prided myself in being free of debilitating injuries, but crabbing brought with it exceptional physical challenges. As a result of handling hundreds of thousands of pounds of king crab that season, the overused knuckles and joints of my index and middle fingers of both hands began to swell.

It was a new and frightening experience, and I awoke one morning to find that my knuckles had puffed to the size of golf balls. Across my palms lay six rows of calluses worn shiny and hard like a turtle's shell. But the calluses were of no use to me, for my swollen knuckles now refused to bend far enough to allow me to perform the simplest tasks, such as gripping the thin walls of my boots and pulling them on. I'm twenty-eight years old, I thought, and I can't even dress myself.

Seeing my predicament on his way out on deck, Bobby uttered but one unemotional word: "Pliers" was all he said.

For the next six days, whenever it came time to pull on my steel-toed Uniroyal boots—size 13½—I'd maneuver a pair of pliers into each hand, pinch the thin walls of each side of a chosen boot, and pull them on, one by one.

Though Bobby's sage advice had solved my boot dilemma, over the next few days the swelling and pain in the knuckles of both hands steadily worsened until it became nearly impossible for me to tie the necessary knots, toss crab, coil, or throw the grappling hook.

At breakfast, Lars noticed the freakish-looking mounds of my knuckles, now twice their normal size. After we'd eaten, he took me aside. "You want to know how to make the swelling go away?" he probed.

I nodded suspiciously. "Well, I would tell you, but you will think I am trying to pull one over on you."

"No I won't, Lars," I insisted. "They're getting bad. Just tell me what to do and I'll do it."

"Okay," he replied, somewhat reassured. "Go into the bathroom and piss on both of them."

I couldn't believe he was serious, and I laughed. "Come on! Fer Christ's sake!"

Lars grew red in the face. "See! Did I not tell you that you would not believe me?" And he stomped off without another word.

By the next day, the swelling in my hands had grown noticeably worse.

Another day and I would no longer be able to do any meaningful work. The throbbing pulse of the pain had also increased. At Bobby's insistence, and risking nothing more than humiliation, I took Lars's advice and made a trip to the bathroom.

As ridiculous as it might seem, over the next twenty-four-hours the swelling in both hands diminished by half, and eventually disappeared almost entirely.

NINE

King crab fishing was plainly the most brutal of all the fisheries. Crab fishermen worked through the coldest and wettest of Alaskan seasons. Pushing 750-pound crab pots about on storm-tossed decks on some of the worst seas on earth, faced with an occupational mortality rate more than twenty times that of coal miners, and driven by crab-crazed skippers to fish for upward of thirty, fifty, sixty, and even eighty hours without sleep or a cooked meal, few deckhands lasted for long.

It is one of the few occupations remaining in the United States in which a laborer's strength, coordination, endurance, skill, and willingness to work not only win a man his job but also play a major role in his surviving it.

Out on the decks of some 150 crab boats scattered all over the Bering Sea, crewmen worked without pause through the dangerous drudgery of four-day tasks, ten-day trips, and a season that threatened to stretch into perpetuity. They ate; they slept, stealing bits of sleep whenever and wherever possible—in a bunk, beneath the galley table, or in the hammocklike webbing of the doorway of a crab pot stacked on deck.

But above all a crewman worked. The best moved straight at it without pause or wonder, spending most of the day bent at the waist, sorting and pitching crabs. On Alaska's rare calm days, the work was merely backbreaking. On rougher days, with the deck pitching steeply and waves exploding over the railing, the sea drenched us repeatedly with icy geysers of salty windblown spray, and the work became pure misery. Several days of it and your body felt as though it had been on the losing end of an all-out brawl.

When the spirit-snuffing hours of darkness closed upon us, we worked on

105

under the hard glare of the deck lights, and usually took a four-hour break for sleep around 2:00 A.M., during which time we were expected to shower, eat, shave, brush our teeth, change our clothes, and, while our skipper slept on, take ninety-minute wheel watches. This commonly meant three hours or less of actual sleep for each crewman.

But there were worse things than three-hour nights. Often we would labor on with no sleep, persisting through the bleak black colors and the strange dreaminess of the hammering repetition and fatigue, determined to last long enough to see the gray spirit-giving light of dawn replace the darkness and mark the beginning of yet another workday.

Commonly, the decks of crab boats are littered with hundreds of crab legs that, despite the best efforts, have broken off in handling. Here and there lay the squashed red carcasses of the adult female and young male crab that had been inadvertently stepped on or crushed beneath a falling crab pot or pot launcher. Scattered everywhere were dozens of claws—some the size of a man's fist—that had broken off when the crab refused to let go of its death grip on the pot's webbing.

In heavy seas, crab pots were commonly washed out of their launchers, and occasionally ended up balanced on the flattened backsides of unlucky crewmen. These crab pots were commonly stored standing upright on crab boat decks. In stormy seas, or whenever the skipper was forced to "run in the trough" to reach other gear, poorly secured crab pots often broke free of their tie-down lines. They tumbled over like giant dominoes. Then they would begin sliding back and forth across the deck, marking or destroying whatever they hit.

A 750-pound crab pot tumbling off the top of a four-tier twenty-foot stack of pots strikes the deck with a velocity of thirty thousand foot-pounds of energy. Human bodies give way instantaneously under such force and are often reduced to a gruesome heap of crushed bone and mangled flesh. A crewman caught in the closing bite of a hydraulic pot launcher will have his leg and pelvic bones broken as easily as one might snap dry straws of uncooked spaghetti.

Every year, ships ran aground, foundered, burned, or collided and sank. But the worst, in terms of loss of life, involved ships that rolled over without warning, often trapping their entire crews inside. Many sank like rocks, taking all hands down with them. Those who did manage to escape the flipping, flooding vessel often found themselves miles from land, without a survival suit in flesh-numbing seawater that rendered them unconscious and drowned them in only minutes.

Gerald Bourgeois, a twenty-nine-year-old king crab fisherman working out of Kodiak, beat the odds, surviving eleven days of almost constant wind, rain, and snow while waiting to be rescued.

Only the week before, Kodiak's Ole Harder, owner and skipper of the fifty-foot *Moonsong,* had spotted a crewman on an island on the south end of Kodiak Island. "You know," Harder told his crew, "that could be a survival suit lying there." Motoring in for a closer view, Harder saw the survival suit move, and exclaimed, "By golly, that's a man!"

Due to shallow water, the *Moonsong* could draw no closer to the shore, but then the crew watched as the man in the survival suit raced forward, dived into the sea, and swam several hundred feet out to them.

Once safely on board the *Moonsong* and recovered, Bourgeois had a harrowing story to tell. During the cold and blustery night of October 3, his crab boat, the *Marion A,* had struck a reef in Geese Channel. It sank in only seconds, spilling all three fishermen aboard into the icy passage. None was wearing a survival suit. As they floundered helplessly in the choppy seas and swift current of Geese Channel, one of the ship's survival suits came drifting through the darkness and Bourgeois snatched it. It had his name written on it.

With the help of a crewmate, Bourgeois pulled on the life-preserving survival suit and stroked toward shore with his friend in tow. But the currents were strong and soon the crewman became too hypothermic to continue.

"We were holding on to one another and he collapsed in my arms," Bourgeois recalls. "The water was rushing all around us in the channel. He had slowed down completely. 'I love you,' he said, and he kissed me on the cheek. I did the same thing back . . . then I knew he was dead. I couldn't do anything. I continued to talk to him. Finally, I had to let him go."

Bourgeois crawled ashore on Aiaktalik Island and rushed to construct a rough shelter to hide out of the murderous piercing chill of the October winds. Marooned without matches and with no way to start a fire, he lived on plant roots, wild celery, and the few mussels he found scattered along the shore. He prayed. He stumbled upon a half-gallon of milk and a bar of chocolate that had washed up on the beach. He consumed both, but over the next eleven days he lost more than 20 percent of his body weight.

Only two days prior to the arrival of the *Moonsong*, USCG helicopter pilot Lt. Jimmy Ng (pronounced Ing) had flown directly over Bourgeois, but a blinding snowstorm had cut his visibility to nearly nothing.

"The sighting would have been the purest luck," the pilot said. "Nobody knew they were even missing!"

Another local Coast Guard pilot told me that Bourgeois's survival was "phenomenal, considering the time of the year and the cold weather. He did all the right things. He built a shelter, ate grass roots, kept out of the wind, and remembered to keep on his survival suit. However, there was one mistake he did make. There was a Coast Guard navigational aid unit there on the island. I guess it never dawned on him to go break the nav-aid light. It would have been reported and fixed at the earliest possible time. Of course he would have been rescued. Also, the light could have been used to signal someone, or [he could have] used the batteries to change the flashing pulse, or just broken into it to draw attention to himself."

The list of injuries and deaths among the fleet around us continued to grow. U.S. Coast Guard records of the season would read like a Marine Corps list of battlefield casualties. And as we learned of each new loss or rescue, we could only cringe at the outcome, wonder at the specifics, and turn back to our work. For a time, however, the news changed things. It cut loose the instincts for survival and sent a new conscientiousness pulsing through our crew, summoning an effort into each knot and step that would not otherwise have been there.

Night and day now, the work aboard the *Williwaw Wind* continued. I learned to keep track of the action on deck by screening the different sounds going on around me. There was the grinding whine of the pot line reeling through the pinch of the hydraulic block, the growl and pop of the line as the boat surged over a wave, the dull thump of a submerged crab pot careening off the ship's bottom, the smack and skid of buoys being tossed on deck, the riflelike explosion of a pot line breaking, and the warning cries of fellow crewmen.

Only rarely during the daylight hours did the stark outline of a passing crab boat show itself on the gray horizon. And to our work-numbed senses, it was a cherished moment of visual relief from the unbroken monotony of the surrounding sea.

At night, our deck lights bore down on us like interrogation lamps, and then a hard black impenetrable darkness would surround us, altered only by the soft, filtered glow of the mast lights of dozens of other crab boats working well beyond the horizon. I found myself longing for those days when I could once again lay my head down and rest without concern for crab, storms, or wheel watches and be assured that no skipper, crewman, or bilge-alarm siren would rouse me.

But there was a season at hand. There were buoys to snag and crab to catch and sleepless twenty-five-hour work vigils to outlast. As the days moved one into another and the crab continued aboard, my mind shifted into a kind of trance, flashing back to pieces of memory, sentimental longings, ancient wrongs, and bits of nostalgia. Worst of all were the involuntary playbacks of some droning television commercial. "How do you spell relief? I spell relief R-O-L-A-I-D-S." Over and over, it would play, my brain helpless to stop it.

This grinding monotony pursued us below decks, too. As one journey flowed into another, the steel cubicle of my stateroom came to feel like a prison cell locked away inside a floating prison ship. "Home" had become a plywood shelf a yard wide and six feet long that was suspended in darkness and padded over with a skid-row mattress. I slept in a cheap cotton sleeping bag worn ragged from use.

Toward the end of the season, I discovered that I was more vulnerable to the problems of isolation and claustrophobia than I would have believed. Living hundreds of miles from civilization, money, too, quickly lost its appeal. Cut off as we were from family and friends and living in the close and confining quarters, the stresses of prolonged periods of work interrupted by a few scant hours of sleep took a spirit toll on everyone.

In the final weeks, we labored on deck as men who could no longer feel the work. The days flooded in over one another, night and day, day and night, passing over us like a strobe-light dream. And my mind began to drift in a fog of physical and emotional fatigue.

Like unthinking oxen, we worked through the drudgery, through the merciless loneliness of living and laboring in a world in which men showed little compassion toward one another, through the tasteless ethnic slights, through the humor and food, through our skipper's critical barbs, through the aching backs and brain-numbing hypothermia, the countless undisclosed injuries, the pitiless nausea of seasickness, the thirty-five-hour days, and the sweaty, grunting, work without end. Amen.

It was during our fourth and last return trip to the Bering Sea that the trouble on board the *Williwaw Wind* began. We were powering ahead through a blinding snowstorm, skirting the black vertical cliffs of Kupreanof Point, when, at approximately 3:00 A.M., Sig shook me awake.

"Your watch, big fella," he whispered.

As he headed back up into the wheelhouse, I climbed into my logger's pants. Shouldering each suspender, I pulled on a sweater, slipped each foot into an open Nike, and clambered up the three flights of stairs leading into the wheelhouse to relieve him. As I entered the room, I could see Sig sitting in the dark. The soft red luminescence of the instrument panel was glowing in his face. He greeted me with a silent glance, and as I approached, he rose from the comfortable perch of the captain's chair, walked to the broad, sloping wood of the chart stand, and flicked on the light. Then, using a compass needle, he pinpointed our ship's position for me on the map.

"We're here now," he explained, "and we will be coming up on the island of Karpa. The skipper, he wants us to cut to the right of her. Except for the 'iron mike' knob [automatic-pilot-system dial], you are to touch nothing! You understand? You got questions, you got problems, you give the skipper a jingle. He's sitting down in the galley playing solitaire with his self. Okay? Oh, and I would not wander far from the jog-stick 'cause the snow is a comin' down pretty thick out there, and the radar and loran is giving us troubles. It is hard to tell what we may come upon. Yah?" Then smiling to himself, he departed.

I snapped off the map light and felt my way to the captain's chair and climbed aboard. Outside, it was a tar black night with sleet and snow flying in steep, thirty-knot flurries. The mast lights had been turned off and we were moving ahead through a darkness that was wet and formless.

Sitting alone in the elevated wheelhouse, high above the deck and passing sea, I could hear the high-pitched squelch of the radio static, the soft belching sound of our diesel engines, and the sandy play of ice chipping at the windows in front of me.

I was staring into the radar screen, studying a suspicious-looking island of green several miles in the distance, when the radar screen blinked off. Then our loran-C machine began pumping out some strange and indecipherable gibberish. When Paul, the young new deckhand we'd picked up in Kodiak, idled into the wheelhouse and I asked him to go below and roust Lars.

I was standing at attention in front of the jog-stick—the lever used to turn the ship in emergencies—when Paul returned.

"The skipper says you should know by now how to read the radar" came the reply. I could feel my pulse quicken. "Help me watch for traffic, will you, Paul?" I asked.

"Watch for *what?*" he shot back as he took my place. "We might as well have our goddamned eyes closed, running through this stuff. I can't see a damned thing!"

As the *Williwaw Wind* tacked blindly through the darkness, I rushed to check out the radar machine. The problem was a burned-out fuse, which I quickly replaced with another of the exact same serial number and strength.

Still, the radar screen showed no signs of life. We must have been well past Kupreanof Point by then, but how and where Karpa Island lay from our present position I could only guess. Equally worrisome was the near-zero visibility. As sea-lanes went, this was a heavily used thoroughfare and the possibility of traffic in the area weighed heavily upon me.

Once again, I sent Paul down to ask Lars to come and help solve the critical problem at hand. Instead of coming to assist me, however, he scolded my messenger. When Paul returned, he reported that he had been told to mind *his* own business.

The weather outside seemed only to worsen, and our ship continued on its blind course, cruising ahead into an opaque and frightening darkness. I could steer the boat but that was all, and as our bow lunged blindly forward, I felt hot blood pumping in my ears.

I was responsible for the safety . . . *no!* . . . the very lives of the crew and I'd been placed in a no-win situation. If I turned on the mast lights, I'd never hear the end of it. Besides, how much good would they do in a blinding snowstorm? How could I possibly know how to fix a radar machine I'd never been allowed to touch?

Peering nervously out one black window and then another, I strained my eyes and senses to the limit in an effort to penetrate the opaque wall of darkness. Sweat formed on my forehead and dampened my palms. Outraged, I spoke aloud: "How could he do this to me?"

With a total of five months sea time, I wasn't yet prepared for this. I kept recalling the mistake I had made on board the *Royal Quarry,* and I felt torn between what I saw as my duty and a nasty confrontation with my skipper. Then I spied a tiny white light hovering far in the starboard distance. In only seconds, I watched it slide by and disappear behind us. Was it a buoy, a boat, or a point of land? I puzzled. And how far off was it? I guessed that we had just passed Kupreanof Point—the intervals between blinks seemed to confirm that—but it might have been a light on a small ship rolling heavily into the troughs. Without a working radar, there was no way to tell for sure.

I heard the skipper laughing aloud in the galley below. Someone must have joined him for a game of cards, I thought, fuming inside.

I decided that by calculating our course, speed, and the time elapsed since our last loran reading, I could attempt a dead reckoning. Still, if an approaching crab boat, tugboat, freighter, or barge lay in our path, I would have no way of knowing it until it was too late. Rock outcroppings were never where they should be. They were always exactly where one found them. Logical estimates of time and location in passages such as these were the tools of fools and only used as backups to the reality of a functioning radar screen.

Forty minutes after first calling for my skipper, I twisted the navigational gyro knob to a heading of 350 degrees and waited. Then came the delayed pull of the ship as it abandoned the old course and swung over to the new. Sick with worry, I could feel the gravity of my decisions gaining momentum with every minute. Too great a starboard swing meant I'd be taking all of us into Stepovak Bay; too early a turn would take us for a short bow-first ride into the rock cliffs of Kupreanof Point. The steering system had hardly finished correcting itself when, for no apparent reason, the radar flashed back on. At that moment, Lars stepped into the wheelhouse.

The stairway ended on the port side of the wheelhouse, and as the skipper walked authoritatively across the wheelhouse floor, I felt terror in my heart but glanced nonchalantly down into the screen to check our course. To my pleasant surprise, my calculations had been nearly perfect. We were moving on a course heading that would place us almost precisely one mile inside of Karpa Island.

"How's it coming?" asked Lars as we exchanged seats. Not waiting for an answer, he added, "How's the radar working? Here, let's take a look." Then, glancing into the radar screen, he said, "This looks okay." His voice sounded surprised.

I made no reply. Lars reached overhead and adjusted the loran calculator and brought it back on line seemingly without effort. Then he strolled leisurely over to the chart board, switched on the tiny light, and, using a diviner, pinpointed us on the map. "Well, how'd it go while I was downstairs?" he asked, trying another tack.

Feeling more confident, I replied, "Oh, the radar just wouldn't work, Lars. And the loran-C computer was down. I checked the radar fuse. It was burnt out, so I replaced it."

"You did vhat?!" he exploded. He seemed to come unglued. "Goddamn it! Show me the fuse! Show me the fuse you took out!"

I opened the drawer and found the spent fuse and the tiny metal box that held copies of the one I'd used as a replacement. Lars grabbed the fuse box from my hands and began to scold angrily. "Which ones did you use? Oh, God! Those aren't the right ones! These are!"

"No, Lars," I patiently tried to explain. "These *are* the right ones! Look! They've got exactly the same serial numbers! They're identical."

If Lars heard me, he gave no notice. "Goddamn it, now! You're going to get me mad! You're going to get me mad!" he screamed, threatening me with a furious ruddy face and poking his finger against my chest as he did so.

Always before, I'd done my best to please the skipper, short of sporting knee pads in the wheelhouse, but I had practiced a humility I hadn't always felt. Mutiny was not something a deckhand took lightly, nor was slowly and deliberately strangling one's skipper and launching his carcass overboard, but in those few moments, both ideas flashed before me.

I'd never before been treated with such disrespect, and my reaction surprised even me. Seething with righteous contempt and my long-smoldering anger, I put my face to his and yelled, a few decibels above his own crude bellow.

"Hey, pal. You're mad? You're mad? You can go straight to your Norwegian hell! I can get mad, too! And you've just won the grand prize! Who the hell do you think you're talking to? Some lackey who's going to lick your boots every time you say *kneel?*"

As I spoke, I inadvertently advanced with my right hand clenched into a fist and the forefinger of my left hand thrusting against his tall chest as I accented each furious word. Faced with such a sudden and unexpected revolt, Lars retreated backward a full stride. Sensing that he was still precariously within my reach, he withdrew farther.

"But, vhat . . . vhat's this?" he said, startled. Then he stood and studied me as if I was something he'd never really taken stock of before. Having seen enough, he fled for the far corner of the wheelhouse. I pursued him, my forefinger poking holes in the air.

"You sit down there on your fat rear playing cards and leave me all alone up here after I send down and ask for your help! And I'm up here without any way in the goddamned *world* to know where we are . . . running wide open in this stuff with no radar! With no running lights! And with orders not to touch a goddamned thing, and all the while without a clue as to what's up ahead! Then you put down the cards long enough to sidle up here and chew my ass out! No sir! Bullshit! That's what I say!"

I took a step closer to make a point, but he interrupted. "You hit a skipper on dah open sea and you vill see vhat trouble you get into!"

Saying not another word, I turned then and stalked from the wheelhouse. I went below and sat down at the galley table. I needed time to cool off. But only moments later, Lars appeared in the galley.

"Young man," he began, "you'd better understand that I knew where we were out here. I've been fishing these waters since before you vas born!"

"The hell you do!" I countered. "You treat people like dogs. The hell with you. You don't know a goddamned thing!"

Lars spun on me, his eyes filled with hate. "You will see how hard it is to come by one of these here jobs!" he said, warning me. "You wait. You just wait and see!" He stomped off a few steps, then turned back toward me. "And you just wait until you are sixty years old! Then ve shall see how much you think you know!" he continued.

"And what, exactly, do you have that I could possibly want, Lars?" I shouted back. "To be sixty years old and loveless and alone? Sounds wonderful!"

"You wait!" he yelled, his fists clenched, his face red and burning with hate. "You just wait, young man! You'll see how easy these jobs is to come by!" Then he stormed from the kitchen and disappeared up the wheelhouse stairs.

So be it, I told myself. A man treated others just as he hoped to be treated. And if he did mean things and his treatment of others was cruel or indifferent, and one day the whole thing blew up in his face, he had no farther to look than himself to place blame.

I had been angry almost beyond words at the time, but as that internal fire of righteousness subsided, I could feel the brief satisfaction I'd known in the

112

wheelhouse quickly evaporate. I'd struck too hard, too brazenly. I'd gone for the throat of a man who was unprepared to do verbal battle. It was a cruel thing to say, my better half argued. And I felt a sense of loss at having become so livid with anger at such a crusty old Norwegian. He was a fine fisherman, and always had been honest with me. Deep down, I knew the man could never be anything other than what he was, and my blowup had served to do nothing other than alienate forever a highly respected Alaskan skipper.

What did I have to complain about? a badgering conscience asked. Hadn't you been fed and housed and given a large cash draw on your first delivery? Had he even once lied to you? Hadn't you been given adequate sleep? And hadn't your skipper not only managed to locate the king crab but also managed to produce a superb market for them? Besides, my gruff and critical skipper was one of the few men I'd ever met who was truly infatuated with the sea. He'd given me a full-share berth on a large Bering Sea king crab boat, set our gear on the crab, and made me more money than I had ever dared hope for in so short a time. Even later, I would realize that the man had made a rare concession in coming to me and trying to explain. But at the time, I had been too hotheaded and inexperienced to understand.

That night in our stateroom, I turned to Bobby. "Why couldn't he just leave me be?" I asked sincerely. "That's all I wanted. He knows I've been pulling my share. Why's he got to act like that?"

"He will now!" he snickered.

"But why couldn't he have before? Look how much money he's making! What is it? Two or three to every one of ours? What does he have to act so unhappy about?"

Bobby smiled and shook his head at my naïveté. "He is the skipper. He is Norwegian." He paused. "You know, you have one hot temper; maybe you get too mad to be at sea for such long times."

The next day at dinner, Bobby cornered me. "Spike! Do not say nothing more to him, not no more at all. He will hate you now till the day he is dead."

"But he was the one who stepped on my toes, Bobby," I argued defensively.

"Yes, but there is nothing you can say that will change him. He will never forget. He will never forgive you. He's got a gun, you know."

Certain that the skipper would no longer want me on board, I told Bobby I'd be getting off in King Cove. Later that day, I was taken aback when Bobby informed me that the skipper wanted me to do no such thing.

"Oh, he does hate you. And he will forever," he explained. "But you have come vith us this far, and you have done good for us, and the skipper figures that you should stay with us and see the season through."

I was astonished that Lars would still want me on his boat, that he would tolerate my services even while he refused to talk to me. It was an unexpected display of fairness that netted me an additional ten thousand dollars in wages, and for which I was very grateful.

TEN

During the last leg of that fourth and last trip to the Bering Sea, Bobby told me that despite the long list of injuries, it was surprising for so late in October that so few crewmen had been killed (knock wood) and that so few crab boats had been lost or damaged.

This was truly surprising considering how hard the fleet had driven itself. Throughout the season, the Bering Sea fleet had caught and delivered 15 million pounds of king crab each and every week and was still going strong when Alaska's Department of Fish and Game announced that the season would close at 6:00 A.M. on October 26.

We consequently stepped up our cruising speed and made record time returning to our pots. But we had no sooner arrived out on the grounds when one of the worst storms ever registered in the Bering Sea rolled in over us.

It was a deep and ever deepening system of low pressure, and as if accelerating into a vortex, the dark ceiling of clouds rolled past overhead in double time, bringing with them a moody presence, a bruise-colored gloom, and humidity.

The marine forecast had predicted "seas very high to thirty feet" and warned of high winds. A deep 962-millibar low was reportedly heading our way. A few hours later, the winds began gusting to sixty knots. It was an easy thing to judge on deck, for it was at this speed that the winds turned savage and began to tear the wave tops from the waves themselves, sending the first of the salty spray tumbling across the face of the ocean.

As the storm developed, surprisingly steep hill-like waves began to move across the sea. With nightfall, a heavy rain began to fall, the winds quickened sharply and great swells arose. Standing out on deck, we waited for the weather to come down, but within the hour the winds fanned to seventy knots, with warning gusts approaching eighty. It proved to be only the beginning. Blowing down out of Russian Siberia, wind gusts approaching one hundred knots soon arrived, having blown unhindered across a thousand miles of open sea.

The storm blew over and around us with the steady, unrelenting pulse of a typhoon. On deck, you could lean into the wind and not fall, and when the speeding molecules of spray struck the unprotected skin of your face, they stung like particles of abrasive rock.

When the barometer fell to an astounding 28.17, we secured all hatch covers, cleared the deck of crab pots and shots of line, and, pointing our bow in the general direction of Amak Island (near the entrance of False Pass), began to idle that way.

Most ships wasted little time breaking for shelter; others remained on the grounds and tried to fish through the mean weather. With each passing hour, the seas grew steeper and more unpredictable, and those hardy or greedy enough to remain on deck found themselves drenched by entire inches of rain and ocean spray, chased by the incomprehensible tonnage of the collapsing waves, and blown literally off their feet.

John Hall, skipper and owner of the highliner crab boat *Provider*, worked his crew well into the storm. In the beginning, he said it wasn't anything he wouldn't have worked in himself. They'd been tacking along, moving into the seas at a precise angle and speed as they continued to pull pots. Then the wind and building seas took on a more dangerous personality. It was "getting ridiculous," Hall recalled. This was "no run-of-the-mill northeasterly!" Some of the waves were beginning to break, so he called his men off the deck.

Hall had been fishing well up in the "compass rose" (waters in the area of the compass configuration found on most standardized maps of the Bering Sea), and with miles and miles to traverse back to the shore at Port Moller or Port Heiden, he recalled how he "just couldn't get out of it. It was some incredible stuff, forty-five footers! But the scary thing was their period. They were coming in real short intervals. You kept your nose pointed into it, hardly moving at all. And your bow was just scooping water. It was no fun, I can tell you."

Scores of crab-boat skippers were faced with the same problem. With great care and concentration, they quartered into the waves and idled toward land. Many, like us, were caught one hundred or more miles to sea. Through the night and all the next day, they pounded into the seas, inching their way back toward Port Heiden or Port Moller, False Pass or Dutch Harbor. There, anchored in the upper reaches of a bay or hiding in the wind shadow of an island, they hoped to find respite from the wind and seas.

That night, all across the Bering Sea, hurricane-force winds tore seawater from the ocean waves and sent translucent walls of spray hurtling across the face of the sea. The roaring spray flashed silver and gray in our mast lights as it smoked horizontally through the night at one hundred knots. It slammed against our wheelhouse walls and the inch-thick Plexiglas of our windows with the resonance and viscosity of blowing sand.

The spray drenched the *Williwaw Wind* in wet, unrelenting surges. It glazed her deck in glistening coats while, overhead, glimmering molecules of ice water paused in the lee of her mast lights, scattering there like the fine weightless crystals of a blizzard snow.

Always before, I'd sensed the overwhelming power and dominance of our ship. She seemed a vessel capable of subduing the sea itself. Yet now, caught in the unforgiving maw of such a storm, she seemed petite and vulnerable.

The waves were wild, heaving, and precipitous. They rolled toward us unpredictably and without quarter. And as they drew near, they more closely resembled mountain ranges than ocean waves. They tossed our ship as if its 145-foot length and 220-ton displacement were inconsequential, and we fought to hold our ground as the canyonlike troughs and steep green slopes swept by us on both sides.

Frightening as the waves were, there was something in their form and power that held me transfixed. I felt at once fearful and exhilarated, for as each new wave ranged up under our hull and swept us skyward, my heart quickened and my skin bristled almost electrically, and I had the sense of coming fully alive.

The *Williwaw Wind*'s mast lights were mounted in a line, high above both her bow and her stern-mounted wheelhouse. On a clear night with calm seas running, their intense beams could illuminate the orange fluorescence of our crab-pot buoy setups several thousand feet out into the darkness. But in the present storm, those stationary beams were rendered almost useless. Mounted rigidly to the masts, they rose and fell along with the ship, vacillating between the converging waves and the close canopy of sky now opaque with spray and darkness.

Once during the night, as we idled into the storm, the trough before us fell sharply away, leaving a yawning pitlike trough below. We were committed to the ride ahead, that was certain. And suddenly, the size of the watery canyon yawning before us struck me. An entire football field could have fitted neatly between the towering wave crests fore and aft. We plummeted ahead, free-falling as if caught in a runaway elevator. Then came a jolting concussion that buckled my knees. It felt as if our ship had impacted solid rock, and I felt frantically for handholds.

Downstairs in the galley, neat stacks of dinner plates floated clear of their foot-deep storm shelves and we heard them shatter across the floor.

In the darkness of the wheelhouse, I saw the tall, lean outline of Lars standing in front of his chair, his feet braced wide, his right hand trembling as he worked the job switch. "Damn this . . . all of this," he said.

There was little time to catch our breath, for where our bow plunged, so our mast light followed, illuminating a massive rich green wall of ocean water, thirty or perhaps forty feet in height, closing directly on us from out of the sleet and darkness ahead.

It was an absolutely mammoth wave, and as it powered toward us, misty veils of spray peeled back from the sharp edge of its crest, vaporous streaks that the wind stripped instantaneously away. The wave closed steadily on us, and the bright green color of the wave's sloping body water glowed almost phosphorescently in the concentrated beams of our mast lights. Across the face of the wave were brilliant white patterns of foam, stretched and flattened by the wind. As if magnified by the light, their every detail stood out in perfect clarity, like bleached membranes beneath a microscope.

We accelerated involuntarily down the broad, sloping backside of the wave and, at the bottom, watched wide-eyed and frozen as our sharp steel bow buried itself in the face of the watery mountain now threatening to drive over us.

The wave collapsed heavily across the full width of our wood-planked deck and exploded with a thundering roar that shook me to the core. With the stomach-shuddering report of a cannon, a two-storied wall of flashing white broth tumbled down the length of our deck. The roaring wave drove headlong into our dark wheelhouse. And, standing with the others in the tomblike darkness, I fought to maintain a masculine image of silent indifference.

It was unimaginable, the burden of sea our ship was now supporting. A single cubic yard of the wet and salty stuff weighed nearly fifteen hundred pounds. Buried beneath such a watery tonnage, our ship shuddered and settled.

Then the *Williwaw Wind*'s buoyance found itself and her bow reared heavily like a submarine surfacing. It drove up out of the sea. Chest-deep seawater began draining off both sides of our badly flooded deck. As her bow soared high, thick green tendrils of ocean water drained from its sharp steel sides, disintegrating instantly in the howling vacuum of wind and spray.

Moving through that wave, our ship plunged into yet another wave trough and began the steady climb up its steep wall. Rising, I could feel the G forces pulling at the flesh of my face, and with our mast lights once again pointing off into the spray-filled darkness, we moved blindly toward the summit. Cresting the wave, our bow extended out into space, the front third of our ship entirely clear of the water. Suddenly and fully exposed to the full force of the one-hundred-knot winds, the ship's entire length hesitated and shuddered.

The *Williwaw Wind* rocked forward in a teeter-totter motion and began to accelerate forward, plunging heavily into the next ocean valley. For several seconds as our stern kicked high, I could feel the sudden loss of power and the thrumming vibration of our propellers as they lifted clear of the sea and spun uselessly in midair.

For an exhausting night and a day, we rose and fell over an endless barrage of these moving mountains of water. Each time we topped a wave, all we dared look for was the next. Always we moved into them and ever they into us.

As we made our way toward shelter, I noticed that fishermen weren't the only ones who suffered. The cold blasts of arctic wind, combined with the monstrous seas, were hard on the seabirds as well, effectively stripping them from the sky.

You could see them, puffins mostly, trying to assume aerodynamic poses as they did their best to escape the wind. They tried to hide on the precipitous slopes of the storm waves or in the very belly of the troughs. They leaned forward with their necks stretched out like weather vanes into the fierce winds, their tropical-looking orange beaks virtually scraping the surface.

Often, the waves they rode would fold forward and roar down upon them. Then, without hesitation, the chunky black and white birds would turn and dive directly into the wave itself. They moved with remarkable speed and agility. Their wings seemed to propel them more efficiently beneath the surface than they did above. These awkward-looking birds, which can dive to depths of several hundred feet, literally flew off through the water. Their small fleeing figures disappeared into the pearlish green depths just an instant before the crushing walls of foam and water exploded down over them. When the wave had passed, they would bob back to the surface, shake their tufted heads free of the water, and stretch out low until the next wave, when they would do battle once again.

ELEVEN

Close by us, caught in the same storm some fifty-five miles north of Amak Island, the newly christened crab boat *Key West* was busy riding out the monstrous seas. Except for the man on watch at the time, the entire crew was fast asleep. They'd gone for six days, sleeping only two hours a night, with only a single hour-long break during the day.

"The rest of the time we fished," recalled Pete Knudsen, one of the crewmen on board at the time. They'd delivered three times already, and with some 225,000 pounds of king crab on board, they were hot "on the crab."

At about 8:00 P.M. the night before, a crewman working on the back deck had been knocked against a crab pot in the rough seas and had chipped a tooth. Seeing a ponderous sea building, the *Key West's* skipper, Harold Pedersen, had called his men off the deck. It was time to quit and await better weather and he began to jog into the seas at about half speed.

That night, the storm around them intensified. The revised marine forecast for those Bering Sea waters called for seventy-five to one-hundred-knot winds and seas building to thirty feet. Like all vessels caught far offshore, the 150-foot *Key West* found its decks awash in seas that broke continually over her stern.

It was during the mid-morning watch that two of the six crab pots stowed on the *Key West's* back deck broke loose and began to slide helter-skelter across the deck. "They were banging around pretty good," said Knudsen, "but it was far too rough to attempt to get near them."

All night long, the storm waves pounded across their stern, filling the ship from railing to railing and leaving her back deck flooded and unable to empty itself before the next wave drove over her. Sometime during the next few hours (a

119

Coast Guard inquiry would determine), the careening crab pots sheared off a twelve-inch vent pipe that led down into the lazaretto compartment below deck. When Roald Pedersen, brother of the skipper, came on duty at approximately 7:40 A.M., he noticed there was an unusually deep amount of wave water rolling across the back deck. Suspicious, he went aft through the engine room and shaft alley to investigate. When Pedersen opened the lazarette door, he saw a solid stream of water flooding in through the vent pipe leading in from the deck overhead. Then he slammed the lazarette door and ran to awaken the skipper and crew.

Awakened with the bad news, skipper Harold Pedersen was up in an instant and moving. When he climbed into the wheelhouse, he could see that the back deck was already under water. Even with the pumps working all out, he was forced to watch helplessly as the stern of his ship settled slowly but irreversibly into the sea. "Wake the crew," he told a crewman.

The hastily awakened crew of the *Key West* grabbed their survival-suit bags and, climbing the stairs into the wheelhouse, donned their suits and waited. Even as they watched, the stern deck continued to sink.

Back aboard the *Williwaw Wind,* we heard Harold Pedersen call for help.

"*Mayday! Mayday!* This is the fishing vessel *Key West!* This is the fishing vessel *Key West!* We're in trouble and taking on water! Our position is fifty-six degrees six minutes north, and one hundred sixty-three degrees six minutes west! *Mayday! Mayday!*"

There was silence in our wheelhouse. I could feel the hair stand up on the back of my neck. "That's the new three-million-dollar boat Harold Pedersen's running!" piped in Bobby finally.

The Coast Guard responded to Pedersen's frantic Mayday by requesting their standard information, but Pedersen wasn't having any of it. "I suppose you will want to know what my boat payments are next!" he spat over the radio.

Convinced of the urgency of the situation, the Coast Guard soon relented, broadcasting over the big set with the news for all the fleet to hear.

"Pom, pom. Pom, pom. Pom, pom. This is the United States Coast Guard *Com-Sta Kodiak;* the United States Coast Guard *Com-Sta Kodiak!* The fishing vessel *Key West* is reportedly in trouble and in need of assistance. All vessels in the area should proceed at once to her current position and stand by to offer assistance." Then he gave the *Key West*'s coordinates and her color, length, and radio frequencies.

By the sounds of the commotion over the big set, the *Key West*'s crew was in need of immediate rescue. The problem lay in the fact that plunging crab boats and their scrambling crews were strung out far and wide across the vast waters of the Bering Sea. No one seemed to be near enough to help. A ship trying to make its way to assist through such a tumultuous storm and dangerous seas could take three or more hours to cover only ten nautical miles. We were more than twenty miles away.

On board the *Key West,* skipper Harold Pedersen watched helplessly as the stern of his shiny new crab boat sank lower and lower into the sea. "Break out the life raft," he instructed his crew. "And get it ready to launch over the side."

Working on top of the wheelhouse, bouncing amid an unrelenting torrent of wind and spray, the crew of the *Key West* accomplished the feat. Returning to the wheelhouse, they stood watching and waiting while out on the back deck giant waves continued to crash in over them.

Pedersen turned to his men and asked: "Is there anything you men want to take with you? If there is, now is the time to go and grab it! Do you have your wallets?" One by one, his crew scurried to their staterooms to fetch their personal belongings. From the wheelhouse, Pedersen could see that the back deck was sinking steadily. Sloping down and disappearing into the sea, its pitch was growing steeper with each passing minute. When his men returned from their individual missions, he ordered them to abandon ship.

Tossed by waves that were at least thirty to fifty feet high, with approximately fifteen feet of ocean water flooding across the stern of their vessel, the crew of the *Key West* waded out into the thigh-deep water sweeping across the high ground of the back deck (nearest the wheelhouse). Then, as they struggled to launch their life raft, even that portion of the stern deck sank from beneath them.

Under those rugged conditions, only three of the crewmen were able to board the raft before a wave washed over them. The rushing water tore the raft loose from the *Key West* and left two men stranded on board the sinking vessel. In the next instant, the wind caught the drifting life raft, whisked it away from the boat, and sent it streaking up and over the steep Bering Sea rollers.

"Jump! Jump!" screamed Pedersen to the two crewmen stranded on the upper deck.

One crewman jumped overboard then and managed to swim to the raft. Crewman Pete Knudsen remained on board. The bow of the *Key West* was pointing skyward at a forty-five-degree angle when Knudsen tied a line to the bow and climbed up and over the railing. As he descended the rope line, the giant steel wedge of the ship's bow leapt and plunged beside him. Terrified, he lowered himself halfway to the water and pushed off the free-falling bow and let go, dropping another ten or so feet into the seas below.

Knudsen surfaced, gasped for air, and frantically stroked free of the leaping, plunging bow metal. Only occasionally visible, the life raft was now drifting quickly away. Over the waves and down through the canyonlike troughs, he managed to keep his eye on the raft. "I swam for it for all I was worth!" Knudsen recalled. "Somehow I made it to the raft and they pulled me aboard."

Now, only skipper Harold Pedersen remained on board the sinking ship. When the flooding had first been discovered, the idea had struck him that he "could gun the huge twin diesel engines full bore and plane the ship on top of the ocean's surface." When that failed, he had thought that if they could just keep the water from pouring in or even slow it substantially, he could "limp in toward the shelter of an island . . . anywhere the seas weren't running so incredibly high, a place where they might have a fighting chance to pump her out and save her."

Pederson did everything he could to keep his ship afloat. "But there was too little time," recalled Pete Knudsen. "Saving the men's lives came first. That was the important thing to him."

With the stern deck lost completely from sight and the wheelhouse floor growing sharply steeper, Pedersen remained in the wheelhouse. Soon, the ship was standing on end, with her entire 150-foot hull extended down into the depths. Only the front half of her wheelhouse and pointed bow remained above the water, pointing skyward and bobbing defiantly in the wind.

A short time later, the Coast Guard radioed Pedersen to ask whether he was sure it was necessary to abandon ship. "No, I just thought I'd go for a little swim," he joked back. "Do you wish to be picked up?" the Coast Guard asked. "Hell, no!" replied Pedersen. "We thought we'd just go out there in the life raft and pull a few more crab pots!"

It wasn't until all power had been lost and seawater was flooding over the top of the main stairway into the wheelhouse itself that Harold Pedersen decided to abandon ship. He leapt overboard and swam away from his boat. When he looked behind him, only ten feet of the bow of the huge vessel remained above water.

Those being tossed about in the bonnet-covered life raft caught only brief glimpses of their skipper. Now and then, they spied him as he bobbed up and over one of the gigantic waves. He was fighting to get to them, but there was no way he could gain on the windblown raft and no way for those inside to slow its departure.

The crab boat *American Eagle* seemed to come out of nowhere. Carefully approaching, they made a pass and threw a shot of crab-pot line toward the life raft. With the raft in tow, their obvious intent had been to drag it up the trawl ramp built into her stern. It was a fine idea, but as the drifting crew of the *Key West* soon discovered, there were complications. As the raft full of crewmen was towed near the *American Eagle*'s stern ramp, her bow dove forward and her stern kicked high, and there to greet the drenched and frightened crew appeared the ship's propeller, fully eight feet in diameter, flashing before them like a giant steel slicer, one that was about to devour them, raft, rope, flesh, teeth, bone, hair, and all.

Chop-chop-chop-chop-chop-chop! sounded the huge propeller as its individual blades slashed through the air. *"No! Nooooo!!!!"* screamed the *Key West* crew, making it plainly understood that the present plan wouldn't do at all.

The crew of the *American Eagle* then towed the raft around to the side up near midships. As the two vessels pitched side by side, they tied a buoy on the end of a piece of line and hung it from the end of the crane. Dangling it over the side—like a playground swing—and timing their lift with the rise and fall of the sea, they off-loaded all seven crewmen one by one.

The moment the last crewman had been lifted aboard, the skipper of the *American Eagle* maneuvered to pick up Harold Pedersen, still drifting over the huge seas. When they swung him aboard, his first words were "Did everybody make it?" His entire crew had made it, he was assured. All had survived. And then he cried.

Several years later while skippering the *Ocean Grace* out of Dutch Harbor, Harold Pedersen and three of his crew lost their lives when the pot-encumbered crab boat he was skippering capsized suddenly and sank. The lone survivor drifted in a life raft for twenty-four hours. Rescued by a passing freighter, he credited Harold Pedersen for pulling him clear of the vessel and saving his life.

TWELVE

We had taken refuge from the storm on the lee side of Amak Island. But while the storm sent us surface-dwelling creatures scurrying for cover, it seemed to have had little effect on the feeding habits of those highly prized bottom-dwelling crustaceans grazing across the floor of the Bering Sea, for when we returned to our gear to pick and stack it, once again our pots rose heavy with king crab, and once again I earned more than one thousand dollars before breakfast.

The weather remained rough but workable, and a bone-chilling wind gave warning of an early winter. A few days later, with our pots stored in shallow water and our tanks crawling with more than 100,000 pounds of king crab, we headed on our final trip back to Kodiak, and home.

As we passed inside Kodiak Island's Ugak Island, Sig and Bobby enticed me to put down my "stupid, dumb book"—Ayn Rand's novel *The Fountainhead*—and join them in a friendly game of stud poker. We were sitting around the galley table, happily anticipating our arrival in Kodiak, when Lars stalked into the room.

Never a man to walk lightly, he looked in my direction and said flatly, "I want you off my ship."

I looked up and smiled. His face was flushed and his eyes refused to meet mine.

"Just get me to the dock, Lars, and make sure my settlement check doesn't bounce," I countered.

The moment was too direct, too painful for my crewmates, who had become friends, and I saw them look down into their laps and away. Though it was his right, Lars had tried to pull a cheap trick. Letting a man go, especially when that man had carried his full weight out on the work deck, was something to be done respectfully and in private.

There was no use in prolonging the scene. I rose and went to my stateroom to pack. I planned to spend a pleasant winter down in Ketchikan, that romantic waterfront city where it had all began. There I would sprawl out and sleep for a month, soaking in the soothing vibrations of Mozart and reading the last unconquered works of Hemingway, Kerouac, and Thomas Wolfe.

I was glad the season was over, and relieved to be getting off the ship. Bobby was still making the midships hawser line fast when I leapt from the ship's hand railing and, with a large pack slung across my back, fled up the dockside ladder.

I'd been fired from my job and ejected from my ship on a cold and barren night, and I had no notion where I'd stay. But now none of that mattered. A deckhand's job would always be as tenuous as a misplaced step or a skipper's whim. Once again, I was a free man—free to find other work, a new job with a new beginning, or work not at all. Nothing could bring down the moment, for I had just pocketed sixty thousand dollars in a single seven-week season.

Shifting the cumbersome bulk of my backpack higher onto my shoulders, I flipped up the collar flap of my Navy pea jacket against the wind. Then I hoofed it across the dock through the crunching white powder to the road that wound along the waterfront behind Cannery Row. The pack felt warm against my back and snug against my shoulders, the weight, of no consequence at all.

Making my way into town, I felt the frosty, invigorating push of winter and, lifting my face to greet it, padded silently over a trackless carpet of snow that lay cold and blue underfoot. Then in the frozen darkness a familiar loneliness swept over me. And I soon accepted it.

Thirty-six hours later, tragedy struck out on the crab grounds. In cold and blustery weather, with the close of the Kodiak area's king crab season only a few days away, the crew of the eighty-foot *Epic* was busy stacking crab pots down off the south end of Kodiak Island when the vessel suddenly rolled over. The U.S. Coast Guard and other crab boats searching on the scene hadn't been able to locate the ship or its crew yet, but the search was under way.

We heard about it ashore, listening to the marine-band radio chatter between the Coast Guard and the searching crab boats. A Mayday call had been received by the Coast Guard and nearby fishermen, but there was no word yet on the number of survivors.

It was earlier that evening, on a bitter cold November sea, with flurries of snow whipping across the ocean, that the trouble aboard the *Epic* began. In an attempt to offset a starboard list, the four crewmen working on her back deck began stacking crab pots on the port side while her skipper simultaneously began transferring fuel to the "high side" tank.

In spite of their efforts, by nightfall the *Epic's* list grew more pronounced. Then suddenly, the boat shifted sharply and began to roll. Crab pots broke loose and accelerated down the steeply sloping deck. The four crewmen were pitched overboard into the icy thirty-nine-degree Gulf of Alaska water.

Crewman Richard Majdic surfaced near the bow and found himself alone and adrift in the close darkness, as steep twelve-foot seas lifted and tossed him. Then he saw the *Epic's* wheelhouse and mast lights slip into the waves.

Without a survival suit, Majdic treaded water in the icy Gulf of Alaska and fought to remain afloat and free of the entangling coils of crab-pot line drifting all around him. Strangely, even in the darkness, the water in the area was illuminated. He soon realized that the light was coming from beneath the surface. Even submerged, the ship's mast lights were still burning.

Then Majdic bumped into skipper Reid Hiner in the water. Seconds later, the two became separated.

"Mayday! Mayday! I'm going over!" was the frantic call the U.S. Coast Guard had picked up in Kodiak. Luckily, the skipper of the *Epic* had communicated to others fishing in the area that he was having problems. It didn't take long to put the two together.

The Coast Guard, as a rule, kept at least one of their 378s (the length and nickname of their huge cruiser ships) patrolling in the area at all times. Each cruiser had a helicopter pad built on its aft deck and carried an H-52 helicopter for exactly such rescue missions.

The 378-foot *Boutwell* was en route back to Kodiak from the Bering Sea when SCC *Com-Sta Kodiak* radioed that the *Epic* had capsized and that there were men in the water. The *Boutwell* was to proceed with all possible urgency to assist.

The *Boutwell's* two gas turbine engines, which together produced some fifteen thousand horsepower, were fully engaged and capable of propelling the 378-foot vessel along at thirty-five knots, about four times the speed of the average crab boat.

As they closed on the search area, one of the crewmen aboard the *Boutwell* wandered back toward the stern to watch the show. He knew that whenever they kicked in their giant gas turbines, an ever-pursuing wall of seawater chased them. It always seemed about to collapse into the immense void just off the stern. Far more spectacular was the action involving the vortex separation where the hull sliced through and divided the sea itself. Leaving the water to rejoin itself, tall, converging walls of water spit skyward in the form of an incredible thirty-foot rooster tail.

When, approximately an hour later, the *Boutwell* had closed to within approximately eighty miles of the area, their H-2 helicopter was launched and soon joined more than a dozen crab boats already searching the area.

Majdic swam around the overturned hull of the *Epic* and ran into Bob Waage and another young crewman. "We need to stay together and remain with the boat!" yelled Majdic. But when he tried to climb aboard the rolling, overturned vessel, a thick layer of algae and the pounding surge of the waves made the hull too slick and perilous to scale.

Then Majdic, Waage, and another crewman became separated from the rest of the crew, and drifted together and waited. Every ten to twelve seconds they would hold their breaths and turn their heads away as another icy November wave came crashing in over them. In between the dunking, gagging waves, Majdic took courage in the sight of the mast lights of a group of crab boats searching through the night for them. There were perhaps a dozen of them and they appeared to be moving in circles. But what if they're only running gear? he worried.

Suddenly, every mast light flickered off. Feeling afraid and alone in the darkness, Majdic and Waage despaired as to what could have happened. Unknown to them, the Coast Guard had instructed the eleven or so crab boats searching in the area to turn off their lights for a time to allow their SAR (Search & Rescue) pilots to fly pattern flights over the area. The currents were running strong in the area, and when the lights came back on, Majdic discovered that he and his two crewmates had drifted away from their ship's rolling, wave-thrashed hull. They were several hundred feet from the overturned *Epic* when the deckhand in Bob Waage's arms began having problems. Seconds later, he fell unconscious. Majdic urged Waage to return to the ship with him. But the heroic young Bob Waage remained steadfast in his refusal to leave his helpless friend.

Only minutes had passed, but clad only in jeans, socks, and a long-john top, Majdic could feel the frightening cold closing fast on him. He knew that without the heat-preserving benefit of a survival suit in such bitterly cold water, few men could remain conscious for more than five minutes. (A man submerged in water loses body heat at twenty-four times the rate he does in air of the same temperature.) He also knew that if he and his crewmates were going to remain alive, it was essential that they swim back to their overturned hull and somehow climb clear of the life-sucking cold of the November seas.

For a few moments, he tried unsuccessfully to drag Bob Waage and his friend back to the boat against the current. Nearly exhausted, Majdic was forced to give up. "Can you still hear me, Bob?" he yelled.

"Yah," replied Waage in a tired voice, adding, "If someone finds you, just tell them I'm going this way."

Majdic received no further response and, leaving his courageous friend, he began making his way back through the choppy seas to the capsized *Epic*. He never saw Waage or his unconscious companion again.

Along the way, Majdic came upon a bobbing set of crab-pot buoys that were not all that far from the ship. He was played out and decided to pause there and rest. As he clung to the heaving, dipping buoys, though, he could feel the hypothermic sleep pulling at his consciousness.

He awoke suddenly. Terrified that he had nearly passed out, he lunged to hold on to the buoy. If you go to sleep, you're dead! he scolded himself. He screamed aloud then. *"Noooooooo!!"* Then he slapped himself. Yet despite these efforts, he felt the deep desire for sleep creep back over him.

In the reflecting mast-light glow from the crab boats searching for him, Majdic noticed the effect the twelve-foot swells were having on the overturned *Epic*. The dark outline of her hull rose on a horizontal plane from her submerged bow to her stern, which jutted into the night. Waves were breaking over its entire length as if it were a reef.

He had one chance, he figured. If he timed his move just right with the swell, he might be able to climb aboard the hull. Even if he did somehow manage to scale the slick, leaping hull in the ten- and twelve-foot seas, he knew the snow squalls (gusting to forty knots) and the freezing spray would eventually kill him. But free of the life-bleeding water, it would take longer. And he prayed for someone to find him before he froze.

Majdic swam toward the overturned ship and worked his way completely around it, only to find that, as before, the hull was too slick to grab on to. It was then that he noticed how the stern moved, how high it soared, and how low it dipped. Dangerous as it was, he realized that if he could somehow climb atop it, he would be carried above the murderous cold of the water, and he decided to make the attempt.

As he paddled near, the hull was a fearsome thing to behold. In the rough and unpredictable seas, tons of unforgiving stern metal rose and fell before him. A six-foot propeller with multiple blades, a thick steel plate of rudder seven feet high, and a soaring, plunging keel passed repeatedly in front of him. When he had closed to within only an arm's reach, he came face-to-face with the jagged and irregular steel figures, objects that promised to crush the life from him at the first opportunity.

But there was no time to contemplate the dangers. He knew that if he remained in the water much longer, he was a dead man.

Moments later, a large wave lifted Majdic into the gusting darkness. The stern of the *Epic* rose and then plunged before him. In the pale orange mast lights of the crab boats searching in the distance, he could make out the black outline of the ship's rudder. As it swept vertically past him, he reached out and managed to snag the top of it with one hand. The soaring rudder jerked him from the water as if he were weightless and suddenly he found himself dangling by a single arm from the huge and rusty plate of steel some ten feet above the sea below.

His hands, feet, and entire body, for that matter, had been severely numbed by that time, and though he would be unable to recall exactly how he managed to scale the slick steel side of the table-sized rudder and keel plate, he soon found himself perched atop it, straddling the sternmost ridge of the hull like a cowboy riding a bucking mount.

Majdic screamed repeatedly for Bob Waage and his friend. But if there was an answer, he could not hear it. In the biting cold of the wind-driven snow and spray, he pulled his shirt up over his head and huddled down, placing his head between his legs. And he worried. What if the air trapped inside this hull were to somehow escape?

When he looked up again (after he did not know how long), he thought he was dreaming, for he found himself staring directly into the blinding glare of a set of crab-boat mast lights only a few fathoms away. It was the *Jeanoah*.

He felt exhausted and elated, and everything seemed suddenly illuminated as in a dream. He could see the scampering crewmen on the deck of the ship idling just out of reach. Then he heard someone scream. It was a deckhand on board the *Jeanoah*. "Stay there! Stay there!" the man insisted. "We'll get a survival suit over to you!"

Clad in a survival suit, a deckhand on board the *Jeanoah* leapt overboard, swam to the *Epic,* and threw Majdic a survival suit. "He seemed so numb," the crewman recalled, "it seemed to take him forever to put on the suit." As he floated in the surging seas beside the capsized ship, the crewman in the water yelled again to Majdic. "Okay, now jump in the water here beside me!"

"No way!" Majdic shouted back. "Bullshit!"

With further coaxing, however, Majdic leapt overboard and swam furiously for the other vessel. The crew of the *Jeanoah* jerked him up and over the side, stripped him of his suit, and took him inside, where they fed him hot food and hot drinks.

As Richard Majdic recovered aboard the *Jeanoah,* the search for Bob Waage and his crewmates continued as the Coast Guard and crab fishermen concentrated their search downwind of the overturned *Epic.*

While the search for the missing crewmen intensified, a team of deep-sea divers was flown out from Kodiak to determine whether anyone had been trapped inside the hull of the overturned vessel, possibly in an air bubble. Divers tapped all along the overturned ship but heard nothing. Then they dived inside the bouncing, capsized *Epic* to inspect its interior. They swam through the staterooms, galley, engine room, pilothouse, and dry-storage areas but found neither survivors nor bodies. On top of the wheelhouse, both life rafts were discovered still mounted securely in place.

Hampered by icy fifty-knot winds and building seas, fellow fishermen, including many friends of Reid Hiner, the *Epic*'s skipper, joined the Coast Guard in the search for possible survivors. Another helicopter and a C-130 Search & Rescue plane equipped with a "midnight sun" (an extremely powerful searchlight) were dispatched out of Kodiak and searched through the night for the other four crewmen, but found nothing.

The U.S. Coast Guard could not allow the eighty-foot hunk of steel to drift and endanger passing vessels, and when efforts to hand the ship over to salvagers got nowhere, crewmen on board the *Boutwell* were ordered to break out their .50-caliber machine gun. As one Coast Guard crewman on the scene recalls, "The gunner's mates had a field day! They must have fired three hundred rounds into the *Epic,* but by that time, all you could see was her stern and the inverted word *Epic* stenciled there."

Back in Kodiak, the headline in a local paper read SIX GOOD MEN LOST HERE IN LESS THAN TWO MONTHS!

PART
THREE

WORKING ON THE EDGE:
FISHING THE ALEUTIAN
ISLANDS AND POINTS
NORTH ABOARD THE
RONDYS

SUMMARY

By the close of the 1978 season, even the most pessimistic of fishing clans could see the mind-boggling profits to be made in the king crab industry. As a result, West Coast shipyards found themselves filled to capacity and backlogged a year in advance with orders for crab boats and equipment.

With the sudden demand for fishing boats 70 to 170 feet in length far exceeding their availability, prices for already-existing vessels rose dramatically.

Unable to find adequate numbers of potential boats for use as crab boats, fishermen flew to the Gulf of Mexico, Texas, Louisiana, and Alabama to invest in everything from oil tenders to the relatively inexpensive Gulf Coast shrimp boats, then took them west, back through the Panama Canal.

Best of all, once in Seattle, fishermen found banks ready to refinance them for far more than they had just paid. With only a few superficial alterations (such as hydraulic pump systems, crab blocks, launchers, and live-tank pumps), one could head north to Alaska to cash in on the bonanza.

Not surprisingly, there was a virtual explosion in shipyard orders for new state-of-the-art Kodiak and Bering Sea king crab boats. With each passing month, new boats were launched, christened, loaded with crab pots, and sent directly north to join the fleet already at work.

Each year, the boats that arrived in Kodiak and Dutch Harbor had the capacity to haul more pots, and to catch, hold, and deliver more crab than the previous year's boats. Spurred on by the spiraling crab prices and 100-million-pound quotas, the Alaskan fleet exploded in a geometric progression of numbers, technology, and profit that could continue only as long as the burgeoning king crab populations around Kodiak and in the Bering Sea continued.

131

ONE

My job aboard the *Williwaw Wind* had ended in November. By December, I'd landed a new job aboard the prestigious, 121-foot crab boat *Rondys* (pronounced ron-dees). I'd become aquainted with *Rondys* owner and skipper Vern Hall while fishing for tanner crab aboard the *Royal Quarry* the year before, and through repeated visits to the *Rondys* (whenever our two ships would deliver in port at the same time) and several wild nights out on the town, I'd come to know his superb deckhands on a first-name basis. Since my tenure aboard the *Williwaw Wind* had come to an end, I'd made a point of keeping in touch with Vern in the wild-card chance that a berth on such a legendary vessel might come open.

I was in Ketchikan when Vern Hall called to tell me that indeed a full-share berth had come open aboard the *Rondys*. "Spike," he began, after I quickly accepted, "some men come down to the boat because they want to go to sea. Their chances of becoming an outstanding crewman are about ninety-five percent. A good worker is so obvious on deck. He's more conscientious than the others. He sticks out like a sore thumb." He paused.

"There are only two things I'll fire a man for," he continued. "One is for falling asleep on watch. The other is for lack of effort. If you're trying and doing a thing wrong, then it'll get explained to you, and you can do it right. If a deckhand does things right on deck, the money will take care of itself. The other skippers will hear about it and he'll naturally gravitate toward the top boats.

"Try to do more than your share. And try to get along with everybody else. If everyone on board will do that, we'll all make more money and things will go smoothly. The *Rondys* has always had just outstanding men right down the line. And I think that's why."

132

With that said, I caught a plane to Anchorage and a connecting flight to Dutch Harbor, where I joined my new crewmates and began my sweaty tenure aboard the *Rondys*. Laboring through a cold, bleak winter, we fished for blue king crab off the Pribilof Islands in the Bering Sea, and for tanner crab off Chirikof and Kodiak Islands in the Gulf of Alaska.

Squatting heavily across the water, the *Rondys* sported a bulky thirty-four-foot beam. Built in 1938 by the U.S. Navy, she was one of the very first boats to be made of welded steel. She had spent World War II plying the waters of Chesapeake Bay out of Norfolk, Virginia, and was being used as a cargo carrier when, in 1967, the Hall family purchased her and began converting her into a crab boat.

The *Rondys* was the antithesis of the stately new ships arriving on the scene in Alaska, with their regal interiors, computer-designed hulls, and paintings of women, warriors, and logo designs tattooed to their steel skins. To glance at her, one not acquainted with her history and reputation might believe she was a tramp boat, or perhaps a carrier of various illegal cargoes.

Cosmetically, at least, the *Rondys* looked to be as primitive a beast as had ever entered a harbor. From bow to stern, broad brownish red streaks of rust ran straight down her sides and into the water from any number of wounds. And the gray paint on her bulwarks and fo'c'sle was dirty, scarred, and peeling.

With the streaks of rust flowing haggardly down her sides, she cut the profile of an older vessel in one of its closing seasons; but like a world-class boxer in tattered clothing, the *Rondys* was entirely up to the task. Her steel joints had been lap-welded, and to protect her against running aground, the Halls had welded a fourteen-inch I beam along the entire length of the keel. Although she could lay no claim to style or beauty, her prowess at sea throughout the previous winter had filled me with a blind prejudice in her favor. Secure, that was the feeling. She powered through rough seas and negotiated the waves with muscular indifference, moving ahead through ordinary seas like a tank idling through fields of wheat.

Her wheelhouse was mounted on a round ferryboat stern. Made of aluminum, it rose more than twenty feet above the water. From this lofty perch, the captain could look down on the deck, which sprawled in front of him for some ninety feet, like a well-lit stage before a choice balcony seat.

Though beaten and scarred, the deck was made of a knotless South American wood called iron bark. Dark brown, almost black in color, the two-by-six planks of it were elevated on steel railings approximately a foot off the slick steel hull. Spaced for drainage, and offering good footing, this wood was expensive, durable, and hard beyond imagination. It was so dense that in water it ironically sank.

Approximately ten feet above the deck, mounted near the base of the front of the wheelhouse at exactly dead center, two black steel booms, each thirty feet long, angled up and out from their mountings at forty-five-degree angles. One pointed to port and one to starboard. Held in place by steel cables, they were rocklike in their rigidity and, with the help of pulleys, could be swung to lift crab pots in place. Viewed from the shore—or another vessel—the booms extending out over the long forward deck made the *Rondys* look like an immense insect with two protruding black antennae feeling their way across the water.

When we were confronted at sea with unexpected and yet essential repairs, the *Rondys* was almost entirely self-contained. Below deck, in her engine room, was a workshop complete with ratchets, socket sets, wrenches, pneumatic tools, hydraulic hoses and fluid, oil filters, nuts, bolts, screws, nails, fuel pumps, a cutting torch and welder, a drill press and grinder, rods, saws, turbochargers, come-alongs, chain falls, tanks of oil, outboard motors, patch kits, gasket kits, glue, clamps, gas, cleaning solvent, drums for discarded engine oil, and tens of thousands of dollars more in parts and preparedness.

The forward wall was a twisting maze of ten-inch pipes, valves and valve levers, and pumps and pump controls with red and green buttons. Positioned on either side of the two large diesel engines ("mains") sat the more petite auxiliary engines, used to generate electricity and provide power for the all-critical hydraulic systems.

The door leading into the fo'c'sle was all the way forward off the main deck. It was a large steel void with upper and lower levels connected by a spiral stairway. Here, an anchor chain was piled and stored, and ran through an eight-inch pipe that rose vertically from that cubicle, passed through the steel ceiling plate overhead, and along the bow deck to the anchor mounted securely on the outermost point of the bow.

At the bottom of the winding stairway, in the damp and musky-smelling space between the live holding tanks and the sweaty hull itself, foodstuffs were stowed. There were cans of beans, peas, asparagus, corn, carrots, peaches, pears, plums, and stews, as well as bottles of catsup and Tabasco sauce, and boxes of pancake mix, paper towels, and cereal.

In between the shelves, well back in the subterranean hold, stood the ship's fresh water tank, barely visible among the piles of coiled shots of line, tens of miles of the stuff, lying in broad piles that were head-deep and deeper.

The uncompromising forces of the sea pose an unending battle of physical attrition on ships at sea. Electrolysis eats away at the hull; rust corrodes paint and creeps over every bare or unpainted spot of steel. Pumps, engines, radios, and radar equipment eventually give way under the ceaseless pounding of ocean waves and salty sea air.

The *Rondys* was long overdue for repairs and a general paint and primer renovation. So in April, after a long strike and a hard-fought, medium-profit tanner crab season spent off the Alitak Bay end of Kodiak Island, we headed south to Newport, Oregon's Yaquina Bay, and home.

Each year, the Hall-owned crab boats such as the *Rondys* and the *Progress* returned to the coastal town of Newport for repairs. Without pay, and with only the promise of a profitable king crab season in September and October, we worked steadily through the months of April, May, June, and July to get the *Rondys* ready for the trip north again—to Alaska.

Going about our separate tasks, we rubbed and scrubbed, scraped and sandblasted away the old, and, using paintbrushes and spray guns, applied the fresh colors of the new.

Using nothing more than hand towels and Formula 409 cleaner, we began by hand-scrubbing every inch of the surface of the engines and shelves and the smoke-stained engine room's walls. We performed minor feats of carpentry and plumbing. We installed a new "fail-safe" toilet. We stripped the iron-bark boards from the foot-high steel stanchions holding them and sandblasted the entire deck. Then we pulled the tank covers and continued the gargantuan task, sandblasting all four of the ship's giant holding tanks, ceiling, floor, and all four walls. Eventually faced with knee-deep piles of sandblasted rock in the tanks, we shoveled the debris into five-gallon buckets and lifted them out of the fourteen-foot-deep holds by hand. Then we primed every inch of the sandblasted surfaces with a sticky gray substance the viscosity of Elmer's glue and, finally, sprayed on an expensive layer of supertough epoxy paint.

We pumped up buoys and painted our logo letters—*RD*—and numbers on each of them. We drilled holes in bait jars, tied knots, cut door and bridle and crab-pot lines into their individual lengths, and melted both ends to prevent unraveling. In the Hall workshop along the waterfront, another team of local handymen were hired to build scores more of the hot-fishing Hall crab pots. On and on the work went.

Over the decades, Newport had gained a reputation for producing some of the finest skippers and crews in all of the United States. Their business partners, skippers, and crews have made tens of millions of dollars fishing in Alaska. Her fishing clan began with Wilburn Hall, the late Teddy "Old Feller" Painter, Sr., Ray Hall, and the Martinson family, and they were followed by men such as Vern Hall, John Hall, Gary Painter, Teddy Painter, Jr., Mike "Spike" Jones, Terry Greenwald, Beanie Robinson, Jack Hill, Ron Eads, Bob Jacobson, Bill Jacobson, and Bill Alwert, Tom, Rex, and Dave Capri, Danny O'Malley, Richard Rose, Charlie Johnson, Rich Wright, Mike Wilson, Jim Wilson, Irish Eddie, Joe Rock—the boys.

Few who know will contest that over the years more fishermen have come out of the local Newport High School than perhaps any other school in the state; and one would have to look no closer than Tom Luther, a former high school teacher (and now a fine local carpenter), for the reason. During some twenty-two years of teaching, he helped literally hundreds of youngsters get a start within the fishing industry.

"It wasn't unusual in the late Seventies," he recalls, "to see a kid right out of high school who was completely untrained go up to Alaska for three or four months and make a hundred thousand dollars. I mean big money!"

Tom Luther went on to explain how, faced with such competition as the Seattle crab-boat fleet, such a small fishing community along the Oregon coast had come to be so successful and to have such a powerful impact on the fishing grounds in Alaska, some two thousand miles to the north.

"This was originally a local fishery," he said. "People lived here and fished just offshore. And there was enough product here to keep them working locally.

But then came the huge opportunities in Alaska. It didn't take long for our local fishermen to gear up and push north.

"The strongest, most common trait that a fisherman has is individualism," continued Luther. "But for some reason, they [this Newport group] bonded. Even though they were competitive, they were determined to help each other. Maybe that exchange of information, the sharing of gear, of knowledge, of boat savvy and ship technology made the big difference.

"I don't know how many hundreds of times over the past few years as boats have prepared to leave all of us showed up and volunteered to help. We shared a tremendous amount of fellowship. Day after day, we would show up and pitch in to help them load up and then see them off; groceries and goods, materials for families who were up north. No charge. We'd just help each other. And the benefits have come back to everyone at one time or another.

"There's probably twenty boats down here right now from Alaska, and the typical amount of money going into each boat to get ready for the upcoming year runs about a hundred thousand dollars. That's not even talking about the numbers of people working on these vessels and spending money in the motels and restaurants and the local businesses."

Another Newport tradition was the impromptu early-morning gathering at Moe's or the Pip Tide for coffee and conversation. This was a strictly informal before-work gathering of local skippers and fishermen. Here men prodded each other, swapped the latest news on a new piece of equipment or newly hired deckhand, and shared bits of information about the goings-on up north in Alaska. It was a time of fellowship, and the names of deckhands who did well at sea, as well as those who had proven themselves unfit, somehow became known, their names forever marked.

"Burn me a steak, Molley!" yelled a tongue-in-cheek skipper one morning as I sat down for breakfast. "Old Billy says he'll pay for it. The kind of money that guy's been raking in, why, good God, the cost of things don't worry him a bit anymore!"

Add chuckles. Rebuttal.

By mid-August, we were nearly ready to roll north again. Often during those last days of repair, I smiled as I recalled the rough-looking, rust-streaked *Rondys* of old. Now, freshly painted and in perfect repair, with her name stenciled in clear white letters across the clean red paint of her bow, she looked a little pretentious, like Attila the Hun in a finely tailored suit.

Then one day in mid-August, we waved goodbye to friends, relatives, and well-wishers and shoved off out of the harbor. The crowd on the docks offered up a mixed bag of emotions. Several women were crying, while others waved solemnly until we were out of sight. Two youngsters ran along the docks in pursuit of our ship, as if to prolong our departure.

But depart we did. With a twenty-foot high, 100,000-pound load of crab pots covering every inch of our deck, we moved west along the waterfront. With Newport drifting by on our starboard, we idled past the condominiums and Moe's Restaurant, past the canneries and curio shops, the Pip Tide and the U.S.

Coast Guard base and, finally, slid under the tall and magnificent gray arch of the Yaquina Bay Bridge. Then, moving between long rows of giant boulders—jetties—we passed the stark white column of a lighthouse rising up out of a weathered stand of spruce trees on the ridge above.

I would miss Newport. There had been months of hard work, and nights spent drinking and dancing at Jake's and the Pip Tide, and waterskiing on Toledo River, and clam digging and bonfires and beach parties along Oregon's incomparable coastline. And there were the sports cars and women. But there were king crab wandering the floor of a watery wilderness several thousand miles to the north, and so with only a few parting looks back, I turned my gaze to the course ahead, and we sailed northwest across the open sea.

TWO

My crewmates on this trip were some of the best in the fleet. I'd come to this conclusion while fishing with them throughout the winter before. We had hunted for giant blue king crab up off the Pribilof Islands in the Bering Sea and had chased tanner crab around Kodiak and Chirikof islands.

With a full head of hair and pearly white teeth, Terry Sampson had a face one might expect to see on the cover of a health magazine. Married, the father of an infant boy, he had been an outstanding high school football player and wrestler. He was working as a hod carrier in the construction business when during the winter of 1975 to 1976 he found himself out of work "with three mouths to feed and house payments to make."

He was living on Vashon Island outside of Seattle when a fisherman friend suggested that he fly north to Dutch Harbor and look for work as a crewman aboard one of the crab boats that delivered to canneries there. "Go up and get yourself a job," his friend encouraged him. "I'm sure you can do it." When Sampson complained that he didn't have enough money to buy himself a plane ticket, his friend bought it for him, took him to the airport, and saw him off.

Terry Sampson arrived in Dutch Harbor out in the Aleutian Islands with five dollars in his pocket. The plane ticket was a one-way fare. It was December and it was snowing. In the now-famous Elbow Room Bar, a skeptical old Aleut informed him, "Son, everyone's going home now. The big king crab season is over."

Most people were quite friendly, Sampson found, and even sympathetic. They joked among themselves in a tone that seemed to say, Well, here's another one.

When Sampson explained what he hoped to do, and the financial predicament he'd gotten himself into, the bartender laughed and shook his head in a kindly manner.

"I've seen you guys come and go a hundred times over the past year," he said. "You don't know what you've gotten yourself into here. Do you have any idea how many young men just like you have come up here looking for work? It's really difficult to walk up and just go to work on the deck of one of these big crabbers, especially with no experience. But I do wish you luck, and I'll keep my ears open for you."

The bartender's warning proved accurate. For the next two weeks, Sampson tramped the docks, boarded crab boats, and talked with skippers. However, with no experience, and with no contacts in the fishery, he found nothing in the way of a crewman's berth.

Finding the determined young man broke, with no real hope of securing work, the local Aleut villagers were quick to befriend Sampson. When he inquired as to where he could sleep, he was taken to an abandoned cabin in Unalaska village and told he could throw his sleeping bag down and live there free until he found work or gave up and went home.

Sampson would never forget their kindness. He began spending his days at the Elbow Room Bar, where he made new friends each day. He kept his ears open for news of possible job openings on the crab boats visiting port. Some of the older longtime residents who frequented the place laughed kindly at the naïve young Sampson and the blind ambition that had carried him there.

"You're going to have a hard time doing it this way, young man," warned one.

But Sampson forced himself to laugh it off. He knew he was in big trouble with no money to return home and not nearly enough food to eat, but he had a wife and a baby boy back home in Washington and he knew they were depending upon him.

Each morning then, in the blustery wind and blowing snow, Sampson would hike through the middle of the tiny village to the Pan Alaska cannery to look for work. Day and night, he moved up and down the dockside ladders, climbing on board crab boats to inquire about work in eye-to-eye fashion. Repeatedly refused, he would move on to the next vessel and the near certainty of yet another rejection.

Sampson had competitors, too, who rushed from boat to boat along the docks. But none arrived before he did each morning. They were young men like he was, ambitious and proud and, like Sampson, they felt ashamed at being out of work. Nothing could shake a man's self-esteem more thoroughly than to be turned down by one skipper after another, to be told to your face that you simply weren't wanted, ten or twenty or thirty times in a single week by fishermen who had seen you, and met you, and weren't impressed.

Sometimes Sampson would sit on the Pan Alaska docks, huddling in the darkness against the bitter wind and drifting snow through an entire night as he waited hopefully and literally for his ship to come in.

Sampson could recall that during his first week in Unalaska, there was a sudden thaw, which turned the streets muddy with slush and melting snow and made the search slow going. By the second week, he recalled "things froze up solid again." It was then that the heater in his cabin ran dry of fuel oil. Unable to afford to replace it, the interior temperature of the cabin dropped well below freezing.

One day, he realized that he had not eaten in more than two days. Taking his predicament to heart, one local Aleut family stepped forward with an offer of dinner. It was the first and only time he would ever eat seal liver, but the hot cooked meal and warm hospitality set off deep longings for home. There were onions and hot baked potatoes and corn bread and milk—a fine meal offered at no expense by villagers who hardly knew him, a meal to be enjoyed under their roof, in a well-lit room, in the warm comfort of a heated cabin.

Nearly three weeks passed before Sampson learned that the fishing vessel *Rondys* was pulling into port. He was in the Elbow Room Bar when he heard its skipper chattering over the bar's CB radio. When the 121-foot "red sled" tied up to the Pan Alaska dock, Sampson was sitting there waiting, bags in hand. He had no idea whether or not they needed a man, but regardless, he was present, ready, and willing.

To Sampson, the *Rondys* looked "absolutely gigantic," yet even before the ship had finished tying up, he leapt from the dock down onto the pile of crab pots stacked on deck. He scampered across them and felt a growing embarrassment at not being able to locate the wheelhouse door. Then he worked his way around the back of the wheelhouse and there on the stern spotted the large steel Navy door, which at that moment flung open, revealing the grouchy figure of a skipper.

"What the hell do you want?" he demanded.

Though startled, Sampson held his ground.

"I'm looking for a job. Would you have anything coming open on this ship?"

Russ Ott was skipper at that time. He'd been pondering a problem when Sampson happened by. He was looking for a crewman to replace one that had just quit. The *Rondys* would fish for blue king crab off the Pribilof Islands during the next month. In the depths of winter, it promised to be a dreary duty.

Ott could see that Sampson, standing six feet tall and weighing approximately 205 pounds, was stout and healthy. He was young, soft-spoken, and had a polite manner about him. He seemed to lack that neurotic edge that made many fine physical specimens virtually impossible to live with once at sea. Lastly, and most important, he was hungry. Ott figured he had little to lose.

Looking him over again from head to foot and up again, Ott said, "Well, son, we've got a couple of men going south for a while, and the boat will be tied up here for a couple of weeks. I can't tell you right now whether we can hire you or not, but I'd stick around if I were you. We're going to need somebody."

After Sampson had left, Ott turned to another *Rondys* crewman who'd been in the room at the time and asked him, "Well, what do you think?"

"I don't know," offered the crewman, "the guy doesn't have any experience.

That kind of goes against him. It's going to really be tough to begin with. It'd be a lot nicer if we could get an experienced man."

"Hell, man," shot back Russ Ott, "he's a warm body! He'll make a good deck ape!"

Though he hadn't been promised the job, Sampson was elated. Too excited to sleep, that night he lay in his sleeping bag in the cold darkness of his one-room cabin and waited for time to pass. Unable to remain still any longer, he rose and dressed by candlelight well before dawn. Then he literally ran through the village of Unalaska to the *Rondys* to confront the skipper to face his final verdict.

"Well, can you tell me yet?" he asked the sleep-swollen face of Russ Ott. "Do I have the job?"

"Yah. Sure. Fine," replied Ott. "Throw your gear on board. I'm headed south. The owner [Vern Hall] will be flying in soon."

Ready to bust with pride, Sampson knew there was something he had to do. Strolling into the Elbow Room Bar that afternoon, he announced the good news.

"Hey! I got a job on board the *Rondys!*" he yelled.

The bartender's startled look said it all. "Well, well," he replied. "Way to go, young man! And such a boat, too! The *Rondys* is a fine crab boat! Congratulations!" he added as he poured Sampson a drink on the house, his voice full of goodwill.

"You may just make yourself some money. But understand," he added, leaning forward in a serious and confiding manner, "it is a very hard boat to work aboard." Then he backed away and threw up his hands. "But enough of that! For now, we drink to your good fortune!" And raising their glasses, the bartender and several of the Aleuts who'd fed and housed and counseled Sampson over the past weeks offered a toast. "To good fishing!" they said in unison as they tossed back their shots.

My other crewmate, youthful Dave Capri, carried himself with a taut, muscular control. At 160 pounds, he was lean and moved on deck with the refined coordination of a highly trained athlete. An exceptional high school wrestler back in Newport, he had qualified to go to the Oregon State Championships two years in a row.

Like his five brothers, Dave had started work as a fisherman trolling for salmon with his father off the Oregon coast while still in grade school. By his early teens, he was earning nearly ten thousand dollars each summer.

While Terry Sampson worked on deck with a quiet power and emotional restraint, Dave was ambitious and excitable, a tireless blur of quick and decisive action. Thoroughly enjoying himself, he set a pace most men could only hope to challenge, and thrived on the life at sea. When most men played out, they slowed markedly on deck, fell asleep at mealtime, or refused to stir from their bunks. Yet when Dave approached the stub end of his wiry tether, he took on all the strained characteristics of a high-test cable, and just *stayed*.

Acting partly on the recommendation of Newport's Tom Luther, Vern Hall had hired Dave only a week after he graduated from high school. "You get paid what you're worth," Vern told him.

"I had his brother Rex on board for a while," Vern recalled. "And I was pretty impressed. I figured if Dave had half the gumption Rex did, I couldn't go far wrong. So I hired him on at a half share. We made one trip, and Dave was by far the best man on the boat. So I shelved the half-share idea and paid him a full share retroactive. The next trip, I raised him a percent; the next trip I raised him another percent. He is one of those guys that sees things that need to be done and then does them without having to be told. He's been with me ever since."

Looking back, Dave Capri told me, "I had things that I wanted. Most of it was a dream, and Alaska was a way of making it all come true. My brother Rex taught me to be prompt, to get out there and get with it, and do the very best I could do."

During Dave's first season at sea, the *Rondys* caught 1.2 million pounds of tanner crab, and Capri made forty thousand dollars. "Forty thousand dollars was pretty heady wages for an eighteen-year-old greenhorn out of high school!" Capri recalled.

THREE

A week after departing Newport, we touched down in Kodiak and then followed the scent of king crab west to King Cove. For the next ten days, we hauled aboard massive loads of crab pots, packed them around Unimak Island, and dumped them in the designated pot-storage area in the Bering Sea.

We were rushing to make final preparations for the opening day of the 1979 fall king crab season when Vern received word over the big VHS radio that a strike had been called by our fishermen representatives in Dutch Harbor. Word spread almost magically all across Alaska, from harbor to harbor and ship to ship.

Once a strike was called, a fisherman had two choices. He could ignore it, go fishing when the opening day arrived, and forever after face the wrath of his peers as a scab; or, in support of those negotiating in Dutch Harbor, he could go tie up in the harbor of his choice and wait along with the rest of the fleet for the strike to be settled. The skippers and crews of the two hundred or so giant crab boats registered to fish in the Bering Sea were encouraged to go to Dutch Harbor to participate in the strike negotiations themselves—one boat, one vote. At the same time, the skippers of more than 350 crab boats fishing around Kodiak Island were asked to gather in Kodiak for strike negotiations of their own.

Short of a Mayday call for help, nothing could unite a group of fishermen faster than a call to strike, for we all knew that solidarity provided our only hope of receiving a fair price. Without hesitation, Vern Hall swung the *Rondys* to port and headed southwest toward Unalaska Island in the Aleutians. Less than a day later, we sailed into Dutch Harbor and joined more than 150 other crab boats gathered there. They were anchored in the bay or tied up in side-by-side lines that sometimes reached out four, six, and even eight boats deep from the side of the cannery docks. Finding every inch of dock space taken, we anchored in Unalaska Bay to await the outcome of the strike.

Ever an independent breed, Alaskan fishermen had always shunned the idea of unions. Yet through the years, they had embraced the collective, non-union stance of fighting for what they believed was a fair-market price for their product. Holding strike negotiations in which every fisherman refused to fish and, regardless of the size of his vessel, could stand, have his say, and then vote on the latest price offer was an effective front—one with which crabless canneries were forced to deal.

We were represented at the strike table by three highly respected, well-known, and knowledgeable Seattle-area fishermen. There was Bart Eaton (skipper and owner of the legendary crab boat *Amatuli*), Kris Poulsen (skipper and owner of the *Bering Sea*), and the much-respected Rudy Petersen. These men quickly found themselves locked in battle with the two highly competent cannery representatives, men who knew the market and knew their trade.

The Japanese always had been the principal buyers of king crab, purchasing up to 90 percent of the annual harvest. When they set an export price, local cannery representatives had to adjust by ratcheting down their dockside price to fishermen. But there was what one might understate as a small conflict—a disparity of roughly $20 million between the price that fishermen were willing to accept for their king crab and the price canneries were willing to pay.

We longed for the seller's market created the year before in Kodiak when poor fishing and few crab around Kodiak threw canneries into a bidding war. And with more king crab showing up around Kodiak Island, millions of pounds of frozen king crab still unsold in industry freezers, and an estimated 100 million pounds of harvestable king crab in the Bering Sea just waiting to be taken, fishermen everywhere knew it was going to be a buyer's market. The best they could hope for was to get the canneries to freeze their dockside delivery price at a fair level. And so the bartering began.

To the cannery representatives, the fishermen, who opened with a bid of $1.18 a pound, were "a bit high," if not out and out greedy, while the vessel owners believed the canneries' opening offer of 93 cents a pound was "a bit low," if not outright miserly. Legal extortion was the way most fishermen saw it. And with a $25 million discrepancy remaining between the fishermen's bid and the canneries' offer, there seemed to be no end in sight to the strike.

When the opening day of the Bering Sea season arrived, the entire Bering Sea fleet and their crews remained unmoved. Often heated and emotional, the meetings lasted all day, often stretching into the night.

"If we make you this offer, if we give you this price, will you go back and recommend it to your people?" the cannery representative would ask. After each offer, Eaton, Poulsen, and Petersen would return to the second-floor meeting room of the Pan Alaska Cannery building and meet with an entire roomful of crab-boat skippers.

Eaton and the others would present the canneries' latest offer, with their personal recommendations to vote against it. After much discussion, the matter would be put to a vote. In an effort to keep the vote secret, a baseball cap was passed. One boat, one vote.

For an entire week, Eaton, Petersen, and Poulsen held out for a better offer. Each time the cannery representatives bid a price, these select fishermen would return with it and present it to their peers. When the processors' offer came in at 98 cents a pound, Eaton and the others figured they could do better. "We think we can get more," stated Eaton in presenting the latest offer. "We think our crab are worth more than that. So, we recommend that you vote against the offer. Are there any questions? If not, then let's pass the hat."

Occasionally, a few of the more impatient skippers in the ranks would vote to accept the canneries' offer and end the strike. "Gentlemen," Bart Eaton announced after one such vote on the third day, "we need a strong vote on this, to support us in our negotiations. We need everyone here to cast their ballot again so we can return to the bargaining table with a unanimous vote."

No one involved in the strike was legally bound to do so. Huge profits could have been plundered from the free and open waters of the Bering Sea by a single scab crab boat. Such boats as the *Rondys* and *Progress,* both owned by the Hall Corporation, and the *Peggy Joe, Marcy J,* and dozens of others within the fleet could have chosen to break away from the pack and deliver to their own giant processing ships anchored off and waiting to process crab up and down the Alaska Peninsula anytime they chose to. Yet they remained loyal to one another and did their part in presenting a united front to the cannery representatives.

But each day the pressure grew, and the fidgeting among the ranks became more pronounced.

"We're not making any money sitting here in port!" said one skipper at a strike meeting on the fourth day. "Our crews are raising hell, getting drunk and fighting."

Another threw in this comment: "Hell, we're burning fuel on our generators just sitting here. The longer we sit, the more we're going to have to make!"

One fisherman wanted to ignore the price altogether. "If they cut the price in half, screw 'em; I'll just catch twice as many!" he said.

Bart Eaton held his head. Some of these men didn't even have the price of their rain gear invested in the ships they ran. Some of them now wanted to do nothing more than grab their man shares and run. A failed strike vote wouldn't affect their earnings in nearly the same way it would the owner's. But Eaton had little doubt that the majority of the men would listen to reason.

"If you guys don't want to strike, well then, let's go fishing," said Bart Eaton finally. "My boat's paid for. Hell, I don't need this. Let's go fishing! But I'm telling you, we may settle for ninety-eight cents a pound today, but next year they're going to be insisting on only ninety cents a pound, or seventy-five. Sooner or later, you're going to have to stand your ground. You're going to have to stand up to them and show them you're unified and that you're willing to fight!"

Strikes, Eaton continued, rarely raised the price fishermen received for their product, but what strikes did do was help maintain what they already had. No one could claim a perfect record in the negotiations. Eaton could remember once when the Bering Sea fleet went on a tanner crab strike. "We struck for a month, and settled for a penny," he said.

When the ballots were again cast, the vote by fishermen was once again unanimously behind Eaton and his fellow negotiators. The strike would continue.

FOUR

As the strike persisted, hundreds of crewmen were left with plenty of time on their hands, and I took full advantage of it. Throughout the long and wet winter months, there were few harsher and bleaker places on earth than the Aleutian Islands. But now, in summer, the brown and barren landscape became a virtual garden spot. It was a time when tender green shoots of sedge grass and bog grass pushed up through the dead brown layer of winterkill and sprouted. In the direct glare of a midday sun, the rich green color of the grassy hills was blinding. And a hike out into the countryside took one through dazzling islands of sun-yellow buttercups and purple lupine flowers blossoming wide in the long summer light.

The strike also allowed time to enjoy the nighttime pleasures of this isolated harbor. One night around midnight, Sampson, Capri, and I climbed into our Zodiac skiff and struck out across the choppy surface of Iliuliuk Harbor to the tiny village of Unalaska for a night of fun and camaraderie at the famous Elbow Room Bar. And, as I huddled against the wind and small-chop spray, I felt a youthful flutter of excitement welling deep in my bowels—a special, almost magical feeling of having arrived at exactly the right place at precisely the right time.

Eventually, we navigated up a narrow channel that extended into the village itself. Not far from the blue domes of the Russian Orthodox Church, we pulled the skiff ashore and headed for the bar. The ancient Aleut village of Unalaska was a tiny berg of a town that sat on a flat stretch of peninsula land. Its austere and irregularly arranged homes faced Unalaska Bay and the Bering Sea beyond. Surrounded on three sides by water (Unalaska Lake, Iliuknak Bay, and Unalaska

147

Bay), and dwarfed by a backdrop of 2,165-foot Piramid Peak and 1,648-foot Newhall Mountain, Unalaska boasted one store, one post office, one church, one liquor store, gravel streets, and a single saloon, the Elbow Room Bar.

With approximately 150 crab boats on strike and anchored off in the harbor, close to 1,000 rowdy, fun-loving, and thirsty fishermen on hand, more than 2,500 cannery workers temporarily laid off by the strike, and only a handful of police officers on payroll, the bar life became wild and unpredictable, with no insanity left out.

There is but one reason to drink, and that is to achieve a state of drunkenness. Liquor is expensive, smells bad, tastes atrocious, and makes one fat and foolish. But there is a time for everything under the sun. And I was feeling religious.

Against the silence of the village itself, from off in the night, we could hear music and primal screams coming from the legendary bar, sounds that grew steadily louder as we approached. We entered through a residential-like front door and found ourselves in a smoke-filled bar that was even more frenzied than I could have imagined. It was a pit of anxiety and greed, and a wild, rambunctious, and drunken crowd composed almost entirely of men filled the place. The majority of the clientele were boot-clad fishermen. They were milling about on the open floor in a delirium of wild fun and unpredictable excitement. Yet what struck me first was the noise. It was as if with the strike under way and the season pending, everyone had something to say, and they seemed determined to express it at the same time.

A shoulder-to-shoulder and four-deep crowd of fishermen was pressing in to order drinks. While I wrestled my way into the competition, Sampson and Capri somehow managed to secure a small table next to the wall. Fifteen minutes later, with eighteen shots (three six-packs) of golden tequila balanced on a platter, I headed back toward our crew on the other side of the pulsating room. "Coming through! Coming through!" I yelled. I arrived at our table feeling pleased that only two shots had been pilfered en route.

Although it was indeed the proper time to throw back a few, it was also a place to keep one's eyes peeled and have one's crewmates close by. I'd learned long ago not to get too sloshed in an unfamiliar land. Here, you lived by your wits. And if, by the merest chance, a pushing, tumbling mass of scrapping fishermen came crashing your way, you rushed to save your drink and sacrificed your table to the violent flood. A man had to be sensible about the commotion.

Throughout the Elbow Room Bar, an electric prebattle excitement hung in the air. The laughter around me sounded slightly hysterical. I decided to join the moment and in quick order downed six consecutive shot glasses full of tequila. The golden flood burned its way down with all the subtleness of liquid pipe cleaner.

From the oil-field swampland of Louisiana, to the boomtown logging camps of Washington State's Olympic Peninsula, to the bars of salmon season Ketchikan, this was the most wide open and unpretentious bar I had ever visited.

But it was more than the bar. It was the times. It was life on the edge. You could feel it, like a tension that needed venting, a tightness in the chest that demanded release. Those king crab were ours for the taking. Why, it was manifest destiny! It was the money of it, the greed of it, the waste and power and self-righteous all-American glory of it!

Then a large and squatty native woman, her body bloated by a changing body metabolism and decades of unaltered drinking, staggered out of the crowd and over to our table.

She was looking at me. "Hey, you. You buy me dink! Bedder buy me dink!" she commanded.

"I got no money!" I whined, hoping to be left alone.

She turned to Sampson then and asked, "Hey, you!" She staggered and steadied herself. "You buy me dink! Yes?"

"Sure I will," he offered in a loud but barely comprehensible voice over the commotion of the crowd. "Here!" he yelled, taking one of his own shots and placing it in front of her. That was the thing about Sampson: He could accept anyone without judgment, false humility, or arrogance. She sat then, and from the outset, I could tell he liked the woman. He'd known her from his first days in Dutch Harbor when he was broke and looking for work.

During the next few hours of raucous goings-on, I met people from states scattered from Portland, Maine, to Portland, Oregon. Youngsters had come from Minnesota, Wisconsin, Alabama, Florida, Massachusetts, and Rhode Island. They'd come from Kansas, New Jersey, Oregon, Washington, Idaho, California, Nebraska, Montana, and Maine.

They wore baseball caps, Uniroyal boots, and sported wool coats, caps, and sweaters. Some wore their buck knives in leather scabbards laced to their belts, while one packed a sixteen-inch Jim Bowie knife in a scabbard strapped to his right calf. Most noticeable of all, however, and most dearly missed, were the women; the ratio that night being a good forty men for every woman.

Throughout the night, discussions raged over the reputations of crab boats, and their skippers and crews. I heard names such as Russ Ott and his crab boat, *Ocean Leader*. There were Kris Poulsen's "sea" boats, *Bering Sea, Arctic Sea*, and *North Sea*. There were Harold Daubenspeck's "dog" boats, *Labrador, Retriever*, and *Bulldog*. There was Bill White's "astrology" ever-expanding fleet, composed of the crab boats *Gemini, Virgo, Taurus, Pisces, Scorpio, Aries*, and *Aquarius*. And there was reverent talk of Francis Miller and his highline crabber *Enterprise*, as well as the *Rondys, Amatuli, Provider, Progress, Peggy Joe, Marcy J, Polar Shell, Shelikof Strait*, and several new floating steel legends such as the *Pacific Apollo*, said to carry an unheard of five-hundred crab pots in a single load. And I heard the names *Northern Aurora, Kona, Aleutian Princess*, and *Northern Endeavor*, as well as *Aleutian Mistress* and *Aleutian Breeze;* just a random, thirty-one-ship cross section of some of the biggest and best of a fleet of some two hundred crab boats.

With the Elbow Room Bar packed day and night with hundreds of patrons, it took either a determined drunk or a sober navigator to make his way to the bathroom. On this particular journey, I found it packed with crewmen. Upon entering, I was struck by sweet-smelling plumes of pot smoke, something that sent my lungs to sputtering.

On my way back to our table, I ran into a scraggy, rough-hewn-looking character I'd seen on the streets of Ketchikan during my logging days several years before. He recognized me as I passed by and yelled, "Hey!" I turned. Without so much as a word of greeting, he began talking.

He said he'd been rigging crab pots. "All the hours I can stand!" he bragged. "I get ten dollars a pot! Finished twenty-nine of 'em today—lines buoys, bridles, door straps, the works!"

Faking envy, I yelled my approval, wished him well, and, turning away, tried to lose myself in the noisy, milling crowd.

"Hey!" he yelled again, closing on me fast.

I cringed at the sound. Damn. I was almost free of the drunken fool.

"Hey!" he said, pulling at my arm.

With a confidential manner, he drew close, and I noticed that he was sweating profusely and that he smelled. Then I caught the acrid whiff of his breath and the complete dilation of his eyes; eyes that bore straight through me without a flicker of humanity.

"You looking?" he asked, glancing suspiciously around. "You looking?"

He was hinting that he knew where there was cocaine for sale, and he tried to coax me out the front door toward the noisy crowds gathered outside. I told him I wasn't interested and began to move away, retreating in the direction of my table.

"Hey!" he called out, drawing nearer once again and grabbing my arm for what I was determined would be absolutely the last time.

My look relayed the message.

Sensing danger, he removed his hand. Then leaning close, he whispered, "Hey, you know any boats that need a good man?"

FIVE

While negotiators for the fleet did battle with cannery representatives over the price per pound of king crab catches, there was another war being waged out on the waters of the Bering Sea. It was a war against poachers, those who had started the season before its official opening, and it was being fought by one man: Alaska State Trooper Lt. Bob Lockman.

At six three and weighing 245 pounds, Bob Lockman was a powerfully built, good-natured man in his late forties who had a reputation for being meticulously honest. Crab thieves and jump-starters knew full well that Lt. Lockman had the unsettling habit of materializing out of nowhere in the roughly 1 million square miles of Bering Sea ocean at exactly the right place and precisely the wrong time.

Where did the state find these guys? they must have wondered. Why wasn't Lockman out skippering a $2 million crab boat of his own? The man simply had to have better things to do with his time. How could such a man shun the lure of wealth and the prestige of owning his own ship? There was no trusting a man that greed couldn't lure.

As a former skipper of his own Alaskan king crab boat, this sea-toughened man knew the islands, inlets, passages, and shorelines of Kodiak Island, the Alaska Peninsula, and the Bering Sea as well as any crab-boat skipper, and far better than most. Fortunately for the fishermen, he went out of his way to perform his work with both fairness and empathy. He knew what it was to stand at the wheel and pull gear for seventy-two hours without a break.

Lockman was gone from home at least as much as the average fisherman, routinely logging 250 days a year apart from his wife and family. There had been

151

entire months when his family had not seen him at all. With multiple buoy setups of an estimated forty thousand king crab pots dotting the surface of the Bering Sea, and only a single Fish and Wildlife boat—the M/V *Vigilant*—to oversee the situation, it seems impossible that such an outnumbered craft could successfully enforce the law of the sea. But as skipper of the seventy-six-foot vessel, Lt. Bob Lockman gave the impossible his very best shot.

Before the *Vigilant* was built, christened, and ordered into the Bering Sea on patrol, "there was nothing to enforce the law of the land," said Lockman. "The state was almost no threat at all to lawbreakers." One season, the state chartered a boat to patrol the region, but of the ships that were caught, the worst that each received was a five-thousand-dollar fine. The fines were almost humorous, when one considered the profits involved.

Well before the September 10 opening of the 1979 Bering Sea king crab season, Lt. Bob Lockman had been at work—and the hunting had been good. "I like to catch the guy who's fishing before the season has opened," he told me. "The guy who's screwing the other fishermen, yah, I like to catch him. It's the paper violations that I don't much enjoy; the licenses and such."

Crab boats caught fishing before the season opens are seized and fined. It's nothing personal, those who know Lockman would tell you. As one skipper said, "If you got busted by Lockman, you had it coming."

Each year before the season opened, Lockman would usually end up seizing two or three boats whose strings of gear had already been baited and put in place. His crew would scramble to dump the crab back overboard. Then, after stacking thirty or thirty-five of the pots on deck, they'd begin cutting the doors off the rest of the crab pots and pile them on deck, rendering the pots useless to catch crab. He would then notify the guilty party and escort him back to Kodiak. It was the same each season, except that in these last few years, the price of king crab had tripled and the numbers of crab had quadrupled, causing the stakes to run into the millions of dollars.

One indiscretion during the 1979 season involved crab pots registered to several ships that hadn't arrived yet in Alaska. Immediately, Lockman and his crew went to work confiscating the illegal gear. He then notified the owners. Even as their ships were approaching Dutch Harbor, state attorneys were negotiating with the boat owner's attorneys. They had hardly touched the dock in Dutch Harbor when they were served seizure papers. In return, the owners handed over a check in the exact amount of the fine: $250,000!

They weren't crooks, Lockman knew. They'd just gotten a little greedy, pushed their luck, and were caught.

Several days before the season had been originally scheduled to begin, Lt. Lockman came upon the gear of a well-known crab boat. The gear was found to be baited and fishing, so the crew of the *Vigilant* went to work stacking what they could. Several days later, Lockman and his crew were still busying themselves emptying the pots as they went and cutting off doors when Lockman heard that boat's skipper on the radio. He was chatting with another skipper.

"Sounds like with the strike and all, you guys are going to have a little time on your hands," put in Lockman.

The skipper came right back over the radio.

"Yah," he offered, "looks like we're going to have a few days off."

His voice was relaxed and friendly.

"Well," shot back Lt. Lockman, "if you've got a few days off, why don't you come on up here and give me a hand. We've been dumping crab out of your gear for the last day and a half!"

Dead silence.

Then a weak voice replied, "Well, where are you?"

Lockman offered him his coordinates, adding, "We'd sure appreciate it if you guys would come up here and give us a hand with this stuff."

There was another long pause, followed by another faint reply. "Roger."

Enough said, thought Lockman.

Instead of coming back and helping out, however, the skipper headed into Dutch Harbor and then the boat just seemed to disappear from Alaskan waters. Rumor had it that another boat met the fleeing crabber on the south end of Kodiak Island and pumped it full of fuel. Now, she was said to have headed due south, straight out across the Gulf of Alaska.

A week later, when Lockman got back to Dutch Harbor, he learned that the crabber was on the run. The ship's panicked response did much to advance Lt. Lockman and his cause, for it implied great power on the side of the law.

When a lawyer from the state attorney general's office in Anchorage called Lockman about the escaping crab boat, Lockman's reaction was characteristic. "Fine. Let 'em go," he said. "It's better than a fine. And it's a whole lot easier. It's a way of saying 'If we catch you fishing before the season, you're going to have to sell your boat to Chile!' "

Lockman and the lawyer from the attorney general's office agreed on the phone to let them run, with the understanding that the boat would be seized and the skipper arrested if they ever returned. Then they notified the owners in Seattle and passed the word along that they could tell their frightened skipper to slow down; he wasn't being pursued. "But if you ever bring that son of a bitch [the boat] back to Alaska, *it's ours!*" said Lockman.

The ship soon returned. The fine was negotiated and the owners wrote out a check for $100,000. But the real cost? By the time the matter had been settled, the crab boat had missed nearly half the season—a $500,000 loss in only two weeks. It was far more than most businessmen could afford. When one added the $100,000 fine and the fuel burned on the trip south, it became a costly, embarrassing fiasco that made greedy and fidgeting fishermen everywhere sit up and swallow hard.

As bad as the legal battles with the state of Alaska were, often the worst part of getting caught was having to face the fishing industry itself. It was a deceitful trick, to cheat on your fellow fishermen, to get a jump on the rest of the fleet that way. The great majority of the fishermen were honest men who had worked long and hard, often for decades, to get what they had. The laws were generally fair to all concerned. Fishermen and canneries alike didn't take kindly to those who cheated.

Occasionally, when the guilty party returned to the Bering Sea, he found that his crab pots had been completely sabotaged, with many of his lines cut, his pot webbing sliced and torn, his bait jars and crab gone, too—all the work of angry fishermen.

Fishermen at sea often communicated to Lt. Lockman the knowledge of cheating fishermen, with an anonymous radio call stating the type and location of the infraction. Without signing on or off, and with no identification, a voice would come over the radio: "Ah, maybe you ought to take a look at the loran reading 47725 by 34240." End of communication.

Lockman has seen it all in his years as skipper of the patrol boat *Vigilant*. One time, several years before, he found himself en route to Dutch Harbor in the Aleutian Islands. He knew the crab season wasn't scheduled to open until the next day, and he didn't want to arrive back in Dutch Harbor before then because he didn't want word to get around as to his exact location.

It was nighttime when he reached the waters offshore of Akutan Bay (not far from Dutch Harbor), and he decided to kill some time. Slowing to a crawl, he began to move ahead at a "no wake" speed. Then he spotted something odd—the lights of a crab boat up ahead. The boat appeared to be working slowly from one point on the water to another as if it was fishing.

Maneuvering without running lights, Lockman maintained a loose contact with the vessel, eventually coming up behind the boat and tailing it for a good distance. It didn't take long to confirm the fact that although the season had yet to open, the vessel was running straight ahead and dumping gear as she went.

Lockman roused his men from their bunks and continued to keep an even pace with the crab boat. Minutes later, he passed the first crab-pot buoy setup. The pot's coiled line could be seen floating and still unraveling on the surface. When he investigated further, he found the pot to be baited and fishing. Stacking the pots on deck, he and his crew began picking them up as fast as they could be gathered.

Finally, the skipper of the illegally fishing crab boat noticed that a ship was tailing him. When he saw that same vessel also was picking up his crab pots, he exploded over the air.

"Hey, you son of a bitch! What the hell do you think you're doing?"

Lockman did not answer, and the irate skipper came charging back to demand an explanation. As he approached, Lockman kept the *Vigilant*'s mast lights shining straight into the ship's wheelhouse windows. Lockman knew it would make the chase and apprehension easier if the errant skipper didn't know who he was until the last moment.

The indignant crab fisherman motored right up to Lockman and the patrol ship before he was able to make out who it was that was lifting his pots.

"And when he saw who it was," recalled Lockman, "he panicked and just took off."

Lockman realized that with a standard six-hundred-horsepower diesel engine, the escaping vessel would have trouble topping ten knots. The *Vigilant*, on the other hand, sported three 12–71 diesel engines and packed a total of two thousand horsepower. At sixteen knots, she could chase down any crab boat in the fleet. As the guilty skipper fled, Lockman and his crew hurried to cut off the remaining crab-pot doors and stack them on deck. When the last door came aboard, Lockman spun the *Vigilant* around, opened up her powerful diesels, and gave chase.

Minutes later, the *Vigilant* pulled alongside the fleeing crab boat. Looking up into the wheelhouse, Lockman could make out the face of the skipper. He was staring straight ahead, frozen at the helm.

Lockman picked up his microphone and hailed him over the ship's loud-speaker. "Hey, partner, are you trying to run away or what?"

The skipper was none too happy when Lockman seized his ship and escorted him back to Dutch Harbor under arrest.

The 1979 preseason proved to be a particularly busy one for Lt. Bob Lockman and his crew. Many fines were eventually levied against crab boats and paid—one for $100,000, and two more for $150,000 each.

Several days before the season was legally set to open, Lockman came upon some four hundred crab pots roughly 190 miles northeast of Dutch Harbor. Although the skipper had not been running them prior to the season, the gear was baited and fishing. There were approximately fifty thousand pounds of king crab already trapped inside the pots. Lockman and his crew once again went to work pulling the gear and returning the crab to the sea. As always, they stacked as many crab pots on deck as the *Vigilant* could carry and then began cutting the doors off the remaining pots. Then Lt. Lockman radioed for a warrant, which was delivered to him in Dutch Harbor.

Lockman knew the skipper personally. He was a "fine fisherman, an honest man with never a past mark against his name." Finding the boat docked in Unalaska Harbor, Lockman crossed over the back decks of several other crab boats (tied side to side) and jumped down on the vessel's back deck. The skipper came out to meet him, and when Lockman gave him the news, the man only looked at him and shrugged his shoulders in philosophic defeat.

Several days later, up on the second floor of the Peter Pan Seafoods Cannery building, the Bering Sea strike negotiations were still under way when something caught the eye of a skipper standing next to a window on the far wall. He turned and leaned closer to the glass. "Well, there goes one of them," he said.

A small, silent crowd of men drew closer to the windows. The skippers were staring down on one of the seized vessels, which was just beginning its seven-hundred-mile journey back to Kodiak, running in the shadow of its escort, the *Vigilant*. "I'm glad it's not me," commented one skipper. "But God almighty, it's a damned tough thing to have happen now!"

Back in Kodiak, a judge would levy a six-figure fine. The message was a warning for every deckhand and skipper in Dutch Harbor: More than the fine, the real price the hapless skipper and his crew would pay was one singularly irreplaceable million-dollar season.

SIX

On the seventh day of negotiations, the local cannery representatives made a $1.01 offer to Bart Eaton and his fellow negotiators. Minutes later, the fishermen's union representatives called another strike meeting.

It was Bart Eaton who addressed the tense throng. "We've been offered one-oh-one a pound for our crab," he began. "And, we're happy to say, the board of directors recommends that you accept it. We're recommending that you vote 'yes' on this ballot."

Several fishermen in the room rocked back in their chairs and slapped their hands together. Short-lived smiles were replaced by the seriousness of the moment as the hat was passed and a new vote was taken. When the ballots were counted again, it was unanimously in favor of settling for the latest offer by the canneries. The strike was over. With a cheer, skippers and crewmen rose from their chairs and shuffled quickly out the door and down the wooden stairway to the dock below.

Outside, a deckhand took off on the run, yelling, "The strike's been settled! The strike's been settled!" And I found myself, along with the others, running to catch my ship.

The news spread faster than I would have thought possible. Soon, dozens of crewmen began to flood onto the docks. They raced here and there, one trying to locate a skipper, another, his crewmate. All along the cannery front, vertical lines of crewmen began descending the dockside ladders to crab boats waiting below, while those who had reached the bottom fanned out to man the tie-up lines, or sprinted across work decks and vaulted over handrailings as they leapt from one ship to the next to get home.

156

In remarkably quick fashion, dozens of columns of jet black diesel smoke shot skyward as the crab-boat engines belched to life. One skipper began signaling his absent crew with his foghorn in excited blasts of three. Others idled impatiently along the waterfront, blowing their air horns and looking for members of their crews.

One skipper, who refused to wait any longer for what he considered the tardy arrival of the skipper of a ship tied next to him, ordered his crew to throw off the clinging ship's lines. Then, leaving the vessel to drift out into the harbor, he headed out the narrow channel for the open sea and the season ahead.

Throughout the strike, Iliuliuk Harbor, between the Dutch Harbor and Unalaska sides, had become jammed with scores of the massive Bering Sea crab boats, but its great vulnerability lay in the fact that it had but one exit. Within minutes of the strike settlement, its calm and protected waters were torn into a lather by a mad horde of overanxious skippers as they maneuvered for position and rushed pell-mell through the narrow channel to gain open water.

The ships dodged one another with snakelike maneuvers. Two vessels that had simultaneously veered from near collisions with other vessels bumped sterns and continued on in their mad pursuit without a word. Cut off by a much smaller, quicker crab boat, one well-known skipper shook his fist, blasted his air horn, and saluted the incompetence of his maneuver with his middle finger.

Anticipating just such a mad departure, our skipper had anchored the *Rondys* outside the mouth of the harbor. So Sampson, Vern, and I hopped aboard our waiting skiff, zipped through the intimidating procession of lumbering ships and tall stern wakes, and were alongside the *Rondys* within only minutes.

We cranked up the main engines, pulled anchor, and soon joined the long, foaming line of massive crab boats. From our wheelhouse, boats scattered toward the horizon. The image on the radar screen called to mind a swarm of bees funneling along the same path. The battle had begun. The spectacle before me was a modern-day version of the Oklahoma Land Grab. The starting gun had sounded, and not until the entire quota was caught and the final whistle blown would anyone dare pause.

As with the Oklahoma Land Grab, there were those fishermen who jumped the gun. We had hardly cleared Priest Rock at the mouth of Unalaska Harbor when several crab boats rolled into town packing 150,000-pound loads of king crab. Ignoring the strike altogether, they'd been hauling king crab aboard their ships for as much as a week by then, fishing as if it were mid-season. These ships (and there were only a few) would deliver the season's first plugged loads—six-figure cargoes of the spiny Alaskan crustaceans.

Legally, the season had been open for more than a week. Besides, we tried to tell ourselves, such deliveries were mere dribblings of the 120-million-pound avalanche of king crab yet to be delivered. It infuriated many fishermen, however, and disgusted scores more. Harvesting crab under a fixed quota system meant that any skipper or crew who refused to recognize the strike had intentionally cheated their fellow fishermen. Yet the $150,000 payoff had proven too much for some.

Our radio hummed with discontented fishermen who cried foul, but short of bearing arms, cutting crab-pot lines, ramming another man's boat, or pitting crew against crew in port, there was little else we could do.

Then came the topper. Only a few days later, with the better part of a trip's catch either delivered or resting in the live-hold tanks of crab boats all across the Bering Sea, Universal Seafoods in Dutch Harbor announced that they would no longer be able to pay the $1.01-a-pound price agreed to during the strike. Furthermore, they announced, they were reducing the dockside price offer to eighty-six cents a pound.

"The dirty bastards!" yelled one irate and anonymous skipper over the radio after hearing the news. "The dirty, low bastards!"

SEVEN

As with the beginning of any fishing season, we began the season faced with the certainty of work and only the hope of profit. On our way out to the gear, the weather took a sharp turn for the worse. Over the span of only a few hours, the speed of the wind increased dramatically and a rough and indecisive sea began to toss us. When the wind began to tear open our raincoats, Capri grabbed a fathom-long piece of ⅜-inch tie-down line (used to tie pots in place) and cinched it tightly around his waist. It was a common way of keeping the wind and cold out, a way of defeating the weather.

Tightly wrapped in rain gear, with his thumbs hooked defiantly over the rope across his belly, he stood upright, facing the wind, and looked out over the barren expanse of dull gray ocean. He seemed lost in thought. With a baseball cap jutting out from the cinched folds of his raincoat hood, and steel-toed boots protruding from beneath his pants, he seemed at once vital and indestructible.

"You look like a samurai warrior!" I chided him, breaking his spell.

"Well, isn't that what this is all about?" he shot back. "To kick ass up here and go home smiling?"

I nodded in agreement. Like Capri, I was nervous about the season ahead and anxious to get going.

We had several hundred pots of unbaited gear already sitting in the water and hundreds of crab pots waiting up the shallow waters of the storage area. With no way to know exactly where the schools of king crab had wandered to, we went to work moving and baiting crab pots, all the while anticipating the moment when, some three days of "soaking" later, we would begin pulling them and lift our first pot.

The weather remained rough. Day and night, a thirty-knot wind gusted across the desolate gray face of the sea. At irregular intervals, small rogue waves smacked the sides of the *Rondys,* sending walls of spray shooting several fathoms into the air. Collapsing heavily across the full width of our ship, the water hissed loudly as it found the deck. The icy, inundating showers of frigid ocean spray rushed down our rain gear in transparent waves and drained from our sleeves in thin, slanting tentacles of seawater.

During those first few days, we reeled aboard long strings of empty crab pots, baited each one with several perforated jars of chopped-up herring, and launched it back over the side. And leaving several hundred of those behind to fish for us, we traveled to the designated pot-storage area and gathered up another entire deckload of crab pots—approximately 125 pots.

With those pots piled high over the entire length of our deck, we were ready to begin our prospecting. This entailed laying out our gear in crab-pot strings of varying length across ocean terrain that the Hall family knew, through decades of fishing, to be historically productive king crab grounds.

On deck, we took great care to make sure that the lines were good, that the crab-pot webbing contained no holes, that the perforated bait jars were filled, and, finally, that the crab pots did not flip upside down when they were launched over the side.

Whenever a crab-boat skipper lays out scatterings of crab pots in search of crab, it is called prospecting. In this process, he dumps over gear at intervals that are twice the normal distance apart. Crisscrossing subsea valleys, or laying gear down the full length of a ten-mile-long trench, a skipper drops crab pots here and there as he tries his luck far and wide, all the while reading his bottom graph as he searches for the soft mud bottom that characteristically is known to support crab populations. Several days later, when our second and third shiploads of crab pots had been picked up, moved, baited, and launched back over the side once again, we lifted our first crab pot. To our delight, it rose heavy with king crab. In fact, throughout the day, our fishing proved excellent, and rich, red-backed king crab appeared in writhing piles in each pot. But on deck, there was little time for celebration. The weather was sloppy and an entire season of work and weather lay ahead.

Throughout the first few days, Vern rarely spoke to us on deck. It was a kind of flattery, because more often than not, he didn't need to.

"Change the sinker line on that pot," he might say, or "Take ten minutes," which we called the "five-minute warning" because we knew that in all probability no more than five minutes would elapse before we would be back out on deck and working again. However, there was usually enough time for a cup of hot cocoa and a peanut butter and honey sandwich, providing that the deck was secure and all preparatory work for the next string had been completed.

Once, he added, "Be a little bit quicker on that hook there," which was an order aimed at me, the hook man. It was my responsibility to hook the pot bridle at the earliest possible moment when it broke clear of the water (after which, I would help control the weight and direction of the pot and guide it aboard).

As skippers went, Vern Hall, with a lifetime of sea experience behind him, was a deckhand's dream. Beginning on his father's *Sea Breeze II,* he'd jigged for albacore off the Oregon-California coast while only a youngster. As his father remembered it, Vern took naturally to the sea. He learned to work. He learned to take pride in what he did. And he enjoyed it. By his second summer at sea, "he was pretty good," recalls Wilburn Hall. "He kicked the work around all right."

During the time when the season was closed, or had not yet begun, Vern, his sisters, and their parents, the whole Hall family, would do gear work down on the Newport docks. There were buoys to paint, webbing to mend, and pots to weld.

In the early 1960s, when Wilburn Hall bought the historic ship *King-n-Wing* and pushed adventurously north into the untapped Dungeness crab grounds around Kodiak Island, Vern chose to go, too.

The Hall value system, one soon came to realize, was "If you can't come by it honestly, it isn't worth having." Anyone who ever worked for the Hall family soon came to realize that. Their values were based on an uncomplicated belief in Christian fundamentals, and in the early days aboard Wilburn Hall's vessel, it was not uncommon for someone aboard ship to lead a prayer out at sea asking for God's blessing of the season ahead.

Each summer, Vern fished with his father aboard the *King-n-Wing* in Alaska. After graduating from high school, he served in the U.S. Navy and spent several years on the coast of Vietnam running an LCI landing craft packing supplies and troops into shore. Then, after the service, Vern returned to Alaska, where he once again went to work for his father aboard the *King-n-Wing.*

In the beginning, the crab-pot limit around Kodiak Island was thirty-five. The fishing ground lay untouched, so the fishing was excellent. Every five to seven days, the Halls would fill up with king crab and deliver to the Wakefield cannery in Whale Pass. They were paid ten cents a pound for their catch. They packed approximately sixty thousand pounds a trip, week after week, quitting only in the spring to return to Newport, Oregon, for ship repairs.

Moving west from Kodiak Island, who could forget the excitement and profits of the pioneer days of the Adak fishery when, in the late 1960s and early 1970s, there was no pot limit and the crab pots came up plugged with three hundred keepers apiece. Vern could recall times when, with only three men on the back deck and the pots constantly filled with crab, they could run only four or five pots an hour, and grew so weary of the process that he found himself praying that the fishing would drop off.

A decade of fishing later found Sampson, Capri, and me on deck, and Vern, now a seasoned veteran in these waters, in the wheelhouse.

The first real disappointment that season came when we began pulling a string of "garbage loads." Inside each of these pots lay a thousand pounds or more of undersized male and female king crab. This more than doubled the work load, and dictated that a half ton of crab be lifted and tossed overboard out of every pot we processed. We knew it would result in a dramatic decline in the rate at which we could turn over our gear, dropping us from some fifteen pots per hour to only eight or ten per hour. Most difficult of all to accept, however, was the dramatic plunge in our profits. Compared to the catches of 50 to 150 king crab per pot we'd been hauling aboard, the present average of 15 to 25 legal male king crab per pot seemed hardly worth the effort. By the beginning of the second day of "garbage fishing," the normally spirited and tenacious nature of our crew had grown a little morose, our attitudes subdued.

Then Vern called down from the wheelhouse. "Well, let's see. I just calculated out that last string for us, and it seems that we averaged only ten pots an hour, which sounds pretty bad until you consider that you've had to toss about four million crabs back overboard. But according to my calculations, you're still earning about one hundred and thirty-eight dollars an hour. Now, do you boys think that's worth your time? Because that's what you're making."

Then it dawned on us, bent and grunting out on the open deck, exactly what we were doing out there and what the work was about. And suddenly, every hour and every crab meant something. Through our skipper's inight, we'd acquired a renewed feeling of accomplishment, a new sense of purpose. Then the gates of adrenaline opened up and the pace on deck quickened sharply.

With the routine on deck established, we bent quickly over each newly dumped pile of king crab. Some fourteen hours of steady work later, Vern called to us. "Secure the deck and come on inside and get warmed up," he said. "And get yourself something to eat. The tide has changed and I have to run to the other end of this string."

Taking advantage of the short break, we went inside and, to a man, sprawled out on our backs across the galley floor, with dripping rain gear, boots, and gloves still on.

With the onslaught of heavy seas, even while working out in the cool, brisk wind, my arch-rival—seasickness—had once again paid me a visit. Back inside, it came over me in a rush of green misery.

I dropped to my hands and knees, crawled under the galley table, and watched as the others scrambled to prepare peanut butter and jelly sandwiches and bucket-sized cereal bowls full of Cheerios and Frosted Flakes drowned in whole milk.

The run to the opposite end of the string took us directly into the trough. As the ride worsened, I lay on the galley floor and rolled from side to side, feeling no more substantial than a jellyfish tossed by the seas.

As I lay there, I inadvertently scanned the familiar scene around me. On the left side of the room, beneath the port-side porthole, sat a torn and moth-eaten couch. It was bolted to the floor and was in such poor condition that no one complained when we flopped down on it wearing our spray-soaked gear. In the middle of the room, against the wall, stood the refrigerator, and across the room, directly in front of that, sat the dining room table. Heavily built, it was a thickly timbered, unadorned slab of hardwood, fronted with a long wooden bench.

On my right, resting against the same wall as the refrigerator, sat the stove. In the far right corner of the room squatted a stainless-steel sink. Following that wall around, one came to the kitchen cupboards. A microwave oven was mounted beneath them.

The door leading to the dressing room and wheelhouse stairway was positioned to the left of the refrigerator. A small sign was nailed above it. The sign read WANTED: HOUSEKEEPER. ROOM, BOARD, & FRINGE BENEFITS. I wondered what it would be like to make a living in a civilized profession in which a man worked only eight hours out of every day, with both feet planted on terra firma.

Approximately five feet above me, I could see portholes cut into the unadorned steel walls, and as I lay there, waves began collapsing over the stern railing. From my sprawled position on the galley floor, I took in an undersea world of green water as the ocean waves flooded our back deck and lapped up and over the galley portholes.

To dispel the ongoing agony of nausea, I hurried into the bathroom, stuck one finger down my throat, and, bending over the toilet bowl, purged my stomach. I returned to the galley on locked knees and flopped down in the big empty couch at the end of the room.

Always one to come to the aid of a suffering crewmate, Dave Capri approached me. I was too taken with my own condition to notice how he held his hands behind him. Suddenly, he shoved a dish in front of my face. The platter was heaped with hot and greasy pork chops covered with a thick layer of brown gravy. Yellow rivers of melting butter were oozing down through it. The wafting grease odor was unavoidable.

"You sick bastard," I told him, shoving the plate away and holding a towel to my mouth.

Exhausted from a never-ending battle with seasickness, rough weather, and constant night-and-day action out on deck, I closed my eyes. In an instant, I felt the weight of sleep descend heavily upon me, and I dozed. Then the muscles of my right forearm spasmed violently and my right hand clamped shut with more force than I could ever muster voluntarily.

No more than a scant ten minutes later, the deep rap and grumble of the ship's diesel engines filled the room. We had come to think of the sudden drop in the engine's rpms as a shouted order. Our skipper, we knew, was slowing in his approach to the next string of crab pots. The sound sent a rush of adrenaline jolting through me and, like the others, I jerked awake and was up and running.

Geysers of seawater exploded off the side of the boat and we ducked our heads as we raced down the hall-like corridor between the wheelhouse wall and the tall stern railing. Not yet fully awake, I was greeted on deck by a cold, brisk slug of buffeting wind. With my eyes watering from the sudden chill and my stomach battling the unending nausea, I stumbled to the bait box and knelt down over the rank-smelling pile of ground-up herring. Then I began filling bait jars as fast as my gloved hands could move.

When I looked up, I noticed Capri. He was staring straight at me—and he was grinning. God, how I loathed that gleeful bastard! I loathed him for his youth. I loathed him for his smooth bronze complexion and his full head of wavy black hair. I loathed him for his beautiful blond girlfriend attending college at

Oregon State University back in Corvallis, who wrote him twenty letters a month, loved only him, and was as intelligent and nice as she was beautiful. And most of all, I loathed the fact that he got stronger at sea, while, for days on end, the hellish misery of nausea gripped my every waking hour.

We had to buck into heavy seas for a few minutes to get to the other end of the crab-pot string. My stomach was threatening to upchuck once again when I saw Capri run forward. I knew what he was up to. Under way in heavy seas, a favorite sport of Dave's was to stand up on the foredeck in front of the fo'c'sle and leap skyward just as the bow of the *Rondys* bucked over the crest of a wave. Then, as the deck fell away and the bow plunged into the trough ahead, he'd be left suspended in air, sometimes as high as ten feet overhead. His eyes would widen with excitement and he'd begin flailing his arms in balancing circles as he struggled to return feetfirst to the deck. He reminded me of someone tossed high by an Eskimo blanket.

This time was no different. As I watched, several of his lofty flights carried him a good eight to ten feet above the deck. He seemed to delight in each new flight. It was as if he found something wonderful and redeeming in escaping the common bonds of earth. And each lofty leap expressed itself on his face in a gleeful expression, as if he was obsessed with levitation. But there was something more in his gaze. It was the condescending look of one peering down on those earthbound spirits that refused to fly.

When he finally tired of his play, he turned to me and stamped his feet and pounded his sinewy chest—a man proud of his good health and enthusiastic about life and the fresh air and all that kind of good crap. He was hopping from foot to foot and popping the thickly gloved fist of one hand into the open palm of the other, like a boxer itching to do battle.

Then he looked into my ashen seasick face and gave me one of his most patronizing smiles. "God, I love it out here," he exclaimed, leaning back and inhaling deeply. "Smell that air!"

"Come over here and let me get a good, old-fashioned stranglehold on you, you arrogant little prick!" I countered, my voice filled with bitter sarcasm.

Capri laughed.

"Hey, Spike!" he said, unmoved by either my misery or hostile tone. "You're not getting seasick again, are yah? Gosh, I hate to see that." He nudged Sampson to draw his attention to our banter. "Remember those greasy pork chops? And the hot melting mayonnaise? Mmmmmmm! Wouldn't a big old greasy plate taste good right now?"

"Come on over here, you sadistic asshole!" I offered in my most benevolent of tones. "Have a taste of this!" As I spoke, I scooped up a handful of spray-soaked herring bits and made a fist. Four columns of mashed herring sludge oozed out from between my fingers like mud from between a child's toes.

EIGHT

As the trip progressed, sleep became a precious commodity. On some crab boats, skippers commonly pushed their men to work for forty, fifty, sixty, and even eighty hours without cooked meals or sleep.

But it was those crewmen who early on in the season ultimately became too tired to care. No longer able to make reasonable judgments, they ended up as casualties in the path of a sliding crab pot, or in the tangle of lines hissing overboard. Even those who came through physically unscathed experienced stupefying bouts of irritation and depression.

In the worst of cases, the endless depression and fatigue aroused anger in a driven crew. Such crewmen soon came to hate the work and loathe themselves for being subject to it. In the sleepless end, they would fight with one another, or make death threats, or snipe at each other like a pack of wild and discontented canines.

Skippers and boat owners alike, who wish to reap the greatest possible benefits from each crewman and hour of sleep, would do well to think of recovery in terms of ninety-minute shots.

Studies have shown that the mind needs approximately ninety minutes to complete one entire sleep cycle. During this cycle, the mind passes through five progressive stages, including the all-important fifth stage, during which rapid eye movement—REM sleep—begins. The quality of a person's sleep is measured in direct proportion to the time spent in this REM level of sleep.

During this time the flow of blood actually doubles, the face and extremities begin to twitch, the eyes begin to move quickly from side to side, and the deepest dreams occur. It is during the time of dreams that the mind adapts emotionally to new situations and solves problems posed during a person's waking hours. In times of severe stress, worry, or intense new learning, large quantities of REM sleep are needed.

It is important to remember, therefore, that the value of any particular period of rest corresponds directly with the amount of time spent in this stage. Anyone who interrupts the sleep cycles prior to the completion of the ninety minutes needed significantly disrupts the mind's ability to refresh itself.

Alternating one man in the bunk while the rest work out on deck is one of the more successful ways skippers can haul more gear aboard ship. Stringing two ninety-minute cycles into three- or even four-and-a-half-hour time-outs is a simple and effective way of maximizing the return for each hour a deckhand spends in his bunk. And remember, the last fifteen minutes are clearly the most important in bringing an exhausted deckhand back to life. So, mentally anyway, three hours of uninterrupted sleep is twice as valuable as two and a half hours, because in the three-hour rest, two complete cycles of REM sleep have been allowed, whereas only one complete cycle of rest is received by the crewman who rests just a half hour less.

The *Rondys* crew was one of the lucky ones. Rarely did we get fewer than four hours of sleep a day. Over the years, Vern had learned that it was far more efficient to rotate men in their bunks and work with men who were halfway rested than to throw everyone into those antiquated marathon sessions of do or die in which no one sleeps. With the help of our fine relief skipper, Lance Russel, spelling us on deck and Vern Hall in the wheelhouse, the crew worked in revolving shifts of twenty hours on and four hours off.

Through the long months of boredom, isolation, and loneliness, a fisherman at sea clung to whatever form of sanity he could muster. At its worst, fatigue and isolation left one too tired or dispirited to care.

Cut off from friends, church, mail, home, civilization, and loved ones, deprived of sunlight and sleep, numbed by spray, snow, and sleet, driven by arctic winds, gorged by bowel-jamming five-minute meals, chased night and day by ice floes and ocean waves, with no land within hundreds of miles and surrounded by a gray and featureless sea in every direction, life offshore that proved to be as much a mental experience as it was a physical one.

In 1860, one Russian sailor eventually stationed in Kodiak wrote home to his family: "The monotony is dreadful. I must confess that when this is over and we leave here, both sound and healthy, I will be ecstatic to meet even Americans. . . . The main problem is that I do not have any news from either of you, and I don't know anything of what is happening in the rest of the world."*

* From *Civil and Savage Encounters: The Worldly Travel Letters of an Imperial Russian Navy Officer, 1860–1861*, by Pavel N. Golovin (Portland, Oreg.: The Press of the Oregon Historical Society, 1983), p. 93.

Understandably, many crewmen soon discover that they are ill-suited to such a life. The months of shipboard confinement, the lack of privacy, and the unending loneliness are enough to make men break down and blubber like children.

At its worst, it grows into a kind of living death. In the darkness and isolation of winter, an illness known as the Aleutian stare overtakes them. The Eskimos call it Arctic hysteria. Miners and trappers have long called it cabin fever. Scientists refer to it as sensory deprivation.

It isn't difficult to identify. The condition is symptomized by glassy eyes, quick irritability, mumbled speech, and irrational behavior. Those overcome by it have been known to rage suddenly over a trifle, or to laugh deliriously over things in which no one else on board can find the slightest humor. Some lose their ability to estimate time or even to recall exactly where (in what ocean) they are fishing. They also may grow paranoid, taking offense at the slightest affront, or become obsessed with thoughts—and occasionally acts—of revenge. An affliction born of isolation and fatigue, it is a depression of the spirit, and it can be dangerous. Time spent away from the sea seems to be the only cure.

Alaska State Trooper Lt. Bob Lockman recalls that in his experience cabin fever wasn't all that uncommon. But rarely did it happen to old-line Alaskans or experienced fishermen. "Most often, they were disturbed people who'd come out of the drug culture, or out of Vietnam," he said.

Among dozens of less dramatic incidents, there was the time a crewman held his skipper at gunpoint. There was the nervous breakdown of a crewman aboard one crab boat, and another aboard a processor, both of which required the removal of the stricken party.

Of the latter incident, Lockman recalled "the man had flipped out completely. He was a big guy and he went berserk! He was running around smashing things with a fire ax." After the man was transferred to the *Vigilant,* Lockman taped him to a chair and he and his crew remained with the man around the clock until he was flown out by bush plane the next day.

Then there was the crewman off the old *Alaska Trader,* who, while fishing out at Adak in the western Aleutians, "flipped out and cut up a couple of guys before being subdued."

Bart Eaton, the skipper and owner of the *Amatuli,* recalls that he once had a crewman "just up and quit" when they delivered a load of crab to a cannery in Dutch Harbor. Initially, it infuriated Eaton and, turning on the crewman, he demanded an explanation.

"Damn it, man! How am I going to find somebody out here whom I can trust to do the job?"

But then Eaton caught the young man's empty-eyed stare. He might as well have been trying to reason with a king crab, for all the emotion and human response he received. Quickly reversing himself, Eaton said to the young man, "Well, now, maybe you're right. Maybe it *is* time for you to get off. Yah, I'm sure of it. I guess it's time."

Though I was never able to confirm it, a crewman in Dutch Harbor told me of an incident involving one afflicted crewman aboard a crab boat fishing in the Bering Sea. He was a greenhorn at the time, and he and the rest of the crew were sitting quietly together at the dinner table, eating their dinner in silence, when his particular youngster suddenly stopped eating. Peering angrily up from his plate, his face was full of accusation.

"Stop that!" was all he said.

"Stop what?" the accused crewman wanted to know.

"You know what!" snapped the irritated youngster.

Amidst puzzled looks, everyone returned to the meal at hand.

A few seconds later, the irritated youngster suddenly reared up and punched the other crewman in the face. A nasty row erupted. Fighting is never condoned at sea, and when the struggling young man was finally restrained and the skipper and crew demanded to know what had caused him to rear up, he pointed to the man he'd struck and in dead seriousness made his accusation clear.

"He was smacking his lips!"

Then there was an "owlish" young gentleman with emotional problems over in Chignik Bay. The skipper who was running things at the time told me the man had a Jekyll and Hyde–type personality and had been working for him aboard his floating salmon processor for several months when he also went on a rampage. "The guy was smart, too," he told me. "Hell, he could speak seven languages, but he just up and decided he would go and kill everybody!" The skipper at the time didn't know what to do. The irate crewman had a knife and a .357 Magnum pistol.

They finally succeeded in wrestling both the knife and the gun away from the crewman and locked him in his room to await the arrival of authorities. The next day, the skipper was lying in his bunk when the crewman appeared again.

The disturbed young man froze in the skipper's doorway and fixed a glassy-eyed stare on him. "Do you see?" he said. "I could have killed you if I had wanted to." Another crewman told me about the suicide of one crewman he knew well. The day before he killed himself, his skipper had handed the man a twenty-thousand-dollar settlement check for the season. "You figure it out!" he challenged.

Three views of the 72-foot *Royal Quarry.*
Top: motoring across Lazy Bay against the
backdrop of Kodiak Island. *Above:* the
back deck, which had to be pieced together
before each trip. *Right:* her back deck
loaded with empty crab pots, each weighing about 750 pounds.

Above: Susey Wagner (foreground) and Spike Walker guide a crab pot as a hydraulic winch pulls it across the deck of the *Royal Quarry* toward the pot launcher or "rack." *Right:* Susey (foreground) and Spike position a pot into the pot launcher on the back deck. *Below:* Launching a pot in heavy seas can be challenging. Here, a crewman aboard the *Amatuli* launches a pot in twelve-foot seas. Photo by Bart Eaton.

Top: If pots stored on deck freeze over, the ice must be chipped away with sledge hammers or baseball bats before the pot can be baited and launched. Frozen pots are a hazard; their excess weight can cause a ship to capsize in rough weather. Photo by Bart Eaton. *Above:* Aboard the *Amatuli,* a crewmember tosses a hook to snag the buoy and line attached to a submerged and "fishing" crab pot. Photo by Bart Eaton. *Right:* After the pot buoy and line are snagged, the hydraulic winch lifts the pot from the ocean floor some 500 to 600 feet below. Here, Spike Walker (right) works to coil the line as the pot is hoisted to the surface. *Royal Quarry* crewman Steve Calhoun is at left.

Above: As the pot breaks the surface, crewmembers guide it over the rail and onto the pot launcher. This pot holds a disappointing number of crab. Photo by Bart Eaton. *Right:* When the catch is more substantial, crewmembers bend to the task of sorting the crab, throwing undersized crab back into the sea and legal-sized crab into holding bins or the live well. Photo by Bart Eaton. *Below right:* After emptying and sorting the crab from one pot, the *Amatuli* moves on to the next buoy and pot, this time in heavy seas. Photo by Bart Eaton.

Above: Spike Walker and Susey Wagner lean on a crab pot jammed with tanner crab. *Right:* Back in Kodiak port, the hatch cover of the *Royal Quarry* is popped off, revealing a ten-foot-deep catch of tanner crab. *Below:* The crew of the *Royal Quarry* (left to right): Spike Walker, Susey Wagner, Mike Jones, and Steve Calhoun.

Top: Back in port, king crab are weighed and then lifted by crane from the live well to the processor or cannery. Photo by Bart Eaton. *Above:* The same live well, after the icy salt water has been pumped out and the cannery workers have gone to work off-loading the crab. Here, their work is almost finished. Photo by Bart Eaton.

Top and above: Two views of Dutch Harbor, with side-by-side rows of crab boats waiting for the season to begin. The huge processing ship *Unisea* can be seen in both photos. Photos by Bart Eaton. *Right:* Spike Walker holds a twelve-pound king crab on board the *Rondys. Below:* The 121-foot *Rondys.* Photo by D. Joyce Hall.

Top: The *Key West.*
Photo by Bart Eaton.
Above and right: Two
photographs of the
Saint Patrick, back in
Kodiak's Dog Bay
Harbor after her di-
saster at sea.

NINE

Aboard the *Rondys,* we worked through the cold drudgery of the hours in a methodical way, moving loads of pots here and there in search of invisible schools of king crab on a muddy ocean bottom some five hundred feet below us. Despite the cold and incessant wind, the rough seas, and the depressing gray pall of the water and sky, the king crab came aboard the *Rondys* at an inspiring rate. Approximately a week after we began working our gear, we delivered 156,000 pounds of king crab into the *All Alaskan* floating processor ship in Port Moller.

Vern and his father and a half dozen other fishermen had purchased the huge old blue and white World War II liberty ship the year before and, at considerable expense, had converted it to a floating processor. The *All Alaskan* could process crab in any bay or inlet with enough water to float it. Now, it sat at anchor at the head of Herendeen Bay with a full crew of some thirty-five laborers, cannery personnel, cooks, and mechanics.

Remaining true to their reputation, the Halls agreed to pay their crews the $1.01 price agreed upon at the strike. My share came to nearly nine thousand dollars. Not bad for a week's hard labor, I thought, choosing to ignore the five months of free shipyard work throughout the spring and summer before.

While excellent fishing and big profits prevailed on board the *Rondys,* many of the scenes aboard crab boats working around us were not so lighthearted. Bruises, contusions, lacerations, and broken bones—every day, more of my fellow deckhands were being injured. In the beginning, the news of death or injury arrived on deck with an unnerving effect. But in time, such news no longer sent shock waves reverberating through my psyche.

169

On board the *Courageous,* a crewman suffered a mild concussion. One crewman working on the deck of the *Arctic Lady* (in Kodiak waters) had his leg broken by a crab pot, while a crewmate sustained a bruised chest. On the *Ironhead,* a twenty-two-year-old youngster had his ankle crunched by a crab pot. A deckhand on board the *Marcy J* was working on deck putting lines away when a freak sea came crashing onto the deck. It lifted and slammed the crewman against a crab pot. Then the wave picked up the lazarette hatch cover and dropped it onto the crewman's left leg, breaking the bone in three places.

We learned that the processing ship *Yardarm Knot* had caught fire over in Dutch Harbor. Fanned by high winds, the fire had burned out of control and trapped and killed one crewman.

Then we heard that a crewman working on board the *Sea Rover* had collapsed. A coworker tried to administer CPR, but the stricken man, we were told, died of a heart attack.

As we were leaving Port Moller, Vern asked to see us in the wheelhouse. "I just wanted to tell you men that this is no time to get sloppy," he began, his hands tucked casually into his front jeans pockets. "Now, I'm not talking about your work on deck. That's been going just fine. I'm talking about the little things that can make or break us. Take our hawser lines [used to tie up to the *All Alaskan*]. There's one tied to the railing up by the fo'c'sle right now."

The deck boss, Terry Sampson, shifted his weight to the opposite foot and looked down. "You guys ought to know that a crab boat not far from us had a hawser line that wasn't properly stored and it spilled out one of their scuppers. Someone left an end hanging over the stern and when the skipper backed down to slow for the first crab pot, it sucked it right down and wrapped itself around their prop.

"Well, it seized up their wheel. Sure, they had twin screws, but that boat is also long and narrow, and they would have had one hell of a time maneuvering without both props. So they had to limp back into Dutch Harbor and waste a good half a load's worth of time in fixing it. That's six figures, one hundred thousand in dollars and cents. So, what I'm saying is, let's not let anything like that happen here. Pay attention, be conscientious, and we'll do all right."

As in a fisherman's dream, sixteen hours later, we arrived back out on our gear to find our crab pots teeming with king crab. The fishing was white-hot, and on deck we worked with excitement and fervor.

In those slack moments when I peered hopefully over the side, I found myself anticipating each crab pot and wondering what it carried. You could see it moving up out of the depths from the muddy bottom some five hundred feet below. It rose into view, a rectangular block of emerald green, as shafts of sunlight penetrated the greenish depths of the sea and reflected off the bleached white undersides of overturned crab inside the pot.

As each steel pot broke the surface, crab three feet wide spilled from its tunnels and tumbled down its sides, and along with the others, I found myself leaning over the side, vying to catch the escaping creatures as they fell.

We dumped the crab from each pot as quickly as they arrived, then bent to the waist-high mound of writhing king crab, sorting out the "keepers" from the "garbage" (the females and undersized male crab) in much the same way miners a few decades before had gleaned the gravels of Nome and the Klondike for nuggets.

Plucking six-pound king crab out of crab pots absolutely blocked full of the gangly, spiny devils filled us with untapped reserves of energy. As I tossed them into the open mouth of our live tank, I felt like a man making a cash deposit in a bank.

On and on, the pots came. As we worked, we unconsciously kept track of what was happening around us by listening for the definitive sounds of labor. There was the smack of buoys arriving on deck, the grinding pop of the line reeling through the pinch of the hydraulic block, the whine of the hook line winching the pot up and over the side, the steel rattle of the crab pot coming to rest in the pot launcher, the wet, smacking sound of footsteps, and the crunch of crab being dumped on deck.

Once, as we ran between strings of gear, we were given another ten-minute reprieve. I was the first to reach the pitcher of fresh water stowed on deck, and hefting it, I downed a quart without a breath. As I leaned back, I felt droplets of sweat racing down the hollow of my back like scampering earwigs. Hurrying inside, I paused in the galley long enough to hold a cereal box overhead. Then I tipped my head back, poured the better part of a bowlful of high-carbohydrate granola into my mouth, and washed it down with a pint of whole-milk chaser.

Through a day and a night, and with few exceptions, each pot rose absolutely teeming with the live and writhing creatures. As the expensive spiny crustaceans inside each pot swung in mass up and over the side, we offered up whistles of approval and greedy exclamations of praise, tempered by the seasoned knowledge that at any time such fine fishing could, apparently without reason, come abruptly to an end.

Sometimes as much as a ton of king crab plugged a single crab pot from corner to corner, and then you could see the writhing weight stretching the pot's webbing and bending the support bars. Under the enormous load, the polyester line would pop and groan in the steel pinch of the hydraulic block, and now and then the 5/8-inch polyester lines gave way without warning, parting suddenly, like old shoelaces.

Such a parting wasn't a casual experience. With a breaking strength of some five thousand pounds, it came in the form of an explosion. The block would leap and the broken lines would whip free, and on deck, you would duck low and scurry for a place to hide, joining the others in an almost unified series of foul curses and vicious oaths. You knew that even as you ran, several thousand dollars' worth of king crab were being carried to the bottom, trapped in a fine steel crab pot that would never again fish for you or yours. There was nothing to be done, though. It was a time to kneel beside the bait box and fill a jar, a time to coil the throwing line and hook, or to set about completing some other small task, all the while trying to forget it ever happened.

The great majority of the time, our crew worked with a simple, unthinking solidarity. I liked and respected my crewmates, and we turned our pots like a trio of meshing gears. Each movement flowed into the next, and each man's efforts became part of a unified whole, until our efforts fused, flowing together as one force.

Now, muscles that had grown society-soft in the comparative ease of ship-yard life began to spring back with the endurance of health and youth. Once again working with my hands in the open sea air, I felt a quiet elation. Scrambling, pushing, pulling, and climbing; buffeted by winds and spattered by spray on a pitching, rolling deck—why, it was enough to make a man feel born again!

On board the *Rondys,* we all had talents. Now, with five seasons of crab fishing behind me, the knots seemed to fall together. A carrick bend, which in the first days aboard the *Royal Quarry* had taken me a dozen attempts and several minutes to complete, now flew together in a stopwatch time of 4.8 seconds. But my biggest asset always had been physical power. Once, during the months we'd spent in the shipyards the summer before, I'd won a sawbuck from Sampson when, on a bet, I placed my right hand across his muscular stomach and stood and pressed his 205-pound body to arm's length overhead, holding him there with one hand.

Always, there was an unspoken rivalry of performance, to do one's assigned tasks with skill and speed and not finish last in doing so. The *Rondys* was a highline crab boat and, seasick or not, I tried to pursue each task like the others, with the hypnotic single-mindedness of one obsessed.

Yet it was here, aboard the ultracompetitive deck of the *Rondys,* that each of us came face-to-face with our own limitations. And now and then, in certain areas, I was forced to deal with the bruised pride of working with coworkers who performed at levels I had not yet attained.

Capri and Sampson were exceptional deckhands, the kind of men who improvised ways of doing a thing and generally accomplished whatever they set out to do. With their unremitting intensity and stellar round-the-clock performances, Capri and Sampson provided ample proof that they were two of the premier deckhands in the entire U.S. fleet. And though there may have been better deckhands, in nearly a decade of fishing, before and after that season, I never met them.

So we worked on deck at breakneck speed, and never did that incredible tide of incoming king crab ebb. It was a pace no deckhand could sustain for long— nor would we have to. For in only twenty-one hours of remarkable fishing, we caught, measured, and live-tanked some 213,000 pounds of legal-sized male king crab.

We were openly relieved when after a half day of traveling and several more tension-packed days of waiting at anchor in Port Moller the boats ahead of us finished with their business and we were finally allowed to deliver our crab. We were elated when the last brailer of king crab was lifted from our deck and hauled aboard the floating processor *All Alaskan.* Our ship had grossed some $213,000. My share came to nearly twelve thousand dollars—all of it earned in less than a day of unbelievable fishing.

TEN

On a vessel such as the *Rondys*, it was never possible to recline comfortably and honestly claim that there was nothing more to do. Through the months and seasons, one learned to ignore the hunger of missed meals, and to do so without harboring adolescent resentments. There was work to be done. On this boat, everything was subordinate to that.

In time, a deckhand learned that if he kept his shoulders and feet dry, he did not get chilled; and even in the worst of icy miseries, he came to know that if he worked hard enough and long enough, he could ward off the stupefying effects of hypothermia and its debilitating stiffness. As mountain climbers say, "When the feet get cold, put on your hat."

An experienced deckhand learned to listen to his body. When his muscles began to quiver and spasm, he gobbled salt and potassium. And he fought the dehydration of megadoses of hard labor by drinking milk, juice, or water, not sugar-based soft drinks loaded with calcium-blocking phosphates.

I had learned through experience that even in the very act of vomiting, a man who felt nearly dead with seasickness could dig down and, if he truly tried, push a 750-pound crab pot across the deck ahead of him.

For weeks on board the *Rondys,* we worked alongside one another and, as a rule, rarely paused or eased in our efforts until the task at hand had been completed. We labored like proud beasts, ignored private aches, and tried not to draw attention to injuries that did not matter. We did so not out of some threat of losing our jobs but out of a sense of pride and duty, and because on board a crab boat the caliber of the *Rondys*, it was expected of you.

173

Now, in mid-season, we grabbed meals between strings of gear and, often as not, sat on the long bench at dinner with our rain-gear pants strewn in piles around our ankles. On our individual runs to the kitchen for extra milk, bread, or utensils, or on longer excursions to the head, we looked ridiculous, like clowns trying to run with their pants down.

This was understandable, though, because while we were at sea and on the gear, there was no time to undress completely; nor was there time to savor the expensive cuts of meat. Our crew could clear the dinner table of an entire six-course meal in only minutes. We ate like ravenous savages who were clearly more concerned with sustenance than eating pleasure. Diary accounts of the time show that one dinner consisted of twenty ounces of top sirloin steak, three glasses of milk, four slices of toast and jelly, peas, applesauce, a large potato buried in butter and sour cream and gravy, and a large bowl of ice cream. My portion, that is.

Each day while working on deck, I experienced fear and pride, anger and greed. Yet it was something more basic than any of these that pulsed through my veins and sent the quadriceps to powering. It arose out of one's instincts, like something primal. And though I occasionally flattered myself by privately laying claim to some superior gift of character or personal courage, secretly I knew that my will to perform arose out of the fear of failure and, most important of all, the will to survive.

The unending stress and loneliness of the life was enough to make a full-grown man break down and blubber like a child with a playground hurt. Even Sampson, our silent, square-jawed, uncomplaining deck boss would, years later, admit to being driven to it. I had, too—but always in the privacy of my bunk, where there were no witnesses. It was a secret place where the pent-up tensions, the loneliness, and the pressures of the life-and-death responsibilities could be eased; a tearful, momentary breakdown forever denied.

Remain long enough on the edge, and you learned the art of lasting. You bulled ahead through sleepless, mind-bending stretches of work, through the emotional binges of crewmen who didn't belong there, and through ice storms that burned your cheeks furnace red. In the darkness of winter, you dealt with the interminable sense of aloneness and week-long bouts of depression that left you drifting in a dull gray world of forced labor and lost reality. You descended through mental and physical levels of fatigue only a military man engaged in direct combat could know. Eventually, you entered that primal state in which you ceased to think in words.

Inside, out of the hypothermic grip of the icy spray and arctic cold, you slept wherever you fell. And in those rare moments of privacy, your prayers spilled out as a mindless, perfunctory habit. For it was sheer existential will that, daily, you were forced to call upon.

For most deckhands, the life of a fisherman was an open exchange. We sacrificed the comfortable life ashore in anticipation of the freedom money would buy us in the future. The worst part of living and working on a ship at sea was the loneliness. As an antidote against it, letters were gathered and saved like precious treasures. Later, in the privacy of one's own bunk, they were read and reread, and pictures were studied and turned until their material took on the soft, pliant texture of cotton cloth.

It was pitiful what a lonely, sleep-starved mind could conjure up. In time, the words and sentences took on meaning well outside anything actually intended. Imprisoned at sea, a crewman often suffered tormented nights of lost sleep over the imagined infidelities of a wife or girlfriend. There, far to the north, forgotten suspicions often grew into open doubt, which eventually gave way to established fact, and then elaborate schemes of entrapment and vengeance were plotted.

It was not an easy show, either, for a wife or a girl left behind in Kodiak or back in states far to the south. The months of loneliness often left her ricocheting between the boredom of virtue and the self-loathing of stepping out on a man she hadn't seen in six months.

Arriving back in port, deckhands often lived out a wild "go to hell" lifestyle. There was the three-day drunk, the three-thousand-dollar night on the town, the reunion with a waiting wife or anxious girlfriend, or the two-hundred-dollar roll in the hay with a pretty face and lusty young female body in a private room at Bee's.

It was there ashore, however, that the worst part of a fisherman's life became reality. It came in the form of bank forms and loan payments, of boat repairs and rising diesel prices, of insurance premiums and questionable marine-surveyor estimates. It manifested itself in the form of battling with one's wife for authority in the home, an authority relinquished during the long months away from home; and in the making of bail for hard-partying crewmen tossed into the clink over what the Kodiak police department viewed as "excessive fun."

In the end, when he'd had enough, he would return again to the sea. There, in that almost exclusively male world, a man could depend on the elemental nature of the life again, and his warriorlike role in it. Laboring against storm winds and ocean waves, he could, once more, become part of the hunt and roam freely across a moody wilderness of sea.

The life of an Alaskan crab fisherman's family also could be brutal in the extreme. The possibility for danger at sea, real and imagined, was almost limitless. Each year, ships and their entire crews vanished at sea without a trace. Ask the hundreds of widows and orphans left behind over the last decade alone. Wives know that at any time they could receive word. Accompanied by the family priest, an Alaska state trooper or city policeman could appear at the door and deliver the somber news.

Sometimes the life or death of a man lost at sea seemed to rest on the mere toss of a coin. Take that stormy mainland night when a crab-boat crewman washed overboard. The ship hardly had reeled around to come back for the man when the skipper and another deckhand became embroiled in a dispute.

"I think I see something on the left!" shouted the crewman.

"I think I see something on the right!" shouted the skipper. But they turned left instead, and found the drifting, drowning man. The toss of a coin.

Those not found often drifted to the bottom, where every gram of flesh, hair, teeth, bone, and cartilage would be scattered in nature's vast and unseen currents as effectively as crematorial ash. And except for those few who knew them, their deaths would pass as anonymously as a lonely arctic breeze.

It is a fact that more men commit suicide over the loss of their jobs than over the death of a wife, child, or loved one. It is not surprising, then, that so few of those conservative masses who find security in a steady paycheck, retirement programs, and long-term employment are lured to an occupation in which a man can find himself hired and fired from two entirely different deckhand jobs in as many weeks.

But by nature, fishermen are a restless, intractable lot. Each season in Alaska, wild and spirited characters secured work in the fleet and when the season was over drew their pay and moved on. They shook hands or offered up an informal salute as, pack slung over a shoulder, they scaled the dock ladder and set off on their way. Some came and went with a complete indifference to security or reputation, and left without one ever knowing their last names.

Many of them were good men, too. Countless times during a given season, you might find yourself trusting such a man with your very life. Yet, having parted company at the season's end, it was not uncommon never to see or talk with him again.

But sometimes, by chance, you would catch wind that your former crewmate was logging in Idaho, or working on a farm in Nebraska, or drilling for oil in Texas, or running a trap line by dog team on a tributary of the Yukon River, or in jail in Seattle, and in your heart you wished him luck and got on with what your days were about.

SUMMARY

Less than a week after our last delivery, we returned to the *All Alaskan* floating processor packing another fine full load of king crab. We were anxious to get our crab off-loaded and get back out to sea and begin harvesting even more crab. We quickly discovered that even though the deliveries were limited only to the owners of the processor such as Oscar Dyson, Harold Jones, and Vern Hall and his father, Wilburn, we still had managed to overtax her sleep-starved crews.

So while the red-hot fishing offshore continued and the limited number of days in the season fled past us, we were forced, like the rest of the fleet, to sit and wait our turn to be off-loaded.

Hurry up and wait became not only our plight but the common condition. Four round-trips from port to the fishing grounds and back again to deliver our plugged live holds of king crab consumed the entire season—a season that lasted only twenty-nine days.

Our catch aboard the *Rondys* had totaled nearly 850,000 pounds. Crew shares for that 1979 season alone came to more than $44,000 per crewman.

The two hundred crab boats that composed the Bering Sea fleet caught and delivered 110 million pounds of live king crab and were paid just under $100 million for them. This catapulted tiny, remote Dutch Harbor into the position of the number-one fishing port in the entire United States for value of product delivered. All three were records attained in a season that proved shorter and more intense than anything fishermen had even seen before. The dreamlike spell of the incredibly rich boom king crab era in Alaska was fully upon us. And we prayed the dream would never end.

ELEVEN

With the fall king crab season over and the 1980 winter tanner crab season yet to begin, I drew my pay and booked a flight south to Portland.

Over the next few weeks away from the northland, my visit in the South 48 turned out to be a sobering one. Everywhere I went, I caught the artificial scent: the best-selling romance novel written in twenty-five days; the anecdote of the five foot one, 140-pound karate student with three months of training who obliterated the six foot five, 270-pound football player; the "all you can eat" diet; the one-minute mothers, fathers, rice.

At a party, I ran into an old high school classmate and asked him what he did for a living.

"I work for the P.U.D.," he bragged, referring to the Public Utilities District.

"You like it there, I take it," I said, noticing his proud smile.

"Oh, hey, it's a perfect job," he offered with the tone of an insider. "You don't do a damned thing."

So many of the faces of my boyhood friends had become filled with a civilized resignation. They talked like people obsessed with the artifacts of the "good life." They had become the success people, the coronary people, the people of toothpaste ads and hair dye, the people with a new car every other year, mortgages, credit cards, two-hundred-dollar Italian shoes, and salon tans.

Joseph Conrad once wrote that a sailor without a ship was as aimless ashore as a log adrift at sea. It was true. Soon I began to miss Kodiak and the amenities of the life I had left. I missed the earthy, unequaled pleasure of working with my hands in the open sea air. I missed the risk and danger, the simple order of life at sea, and, most of all, the strange solitude of places remote.

As my longing for Alaska increased, my intolerance for the world to which I had returned grew and grew. Danger in this world had been minimized to a point where everyday decisions carried comparatively little gravity. Now, I found myself actually longing for the wearisome demands once placed upon me, the everyday heroics, the concentrated attentiveness required in stormy seas, the occasional violence, the prideful demonstrations of physical and mental endurance. I missed the on-deck labor, where a man's skill could be applied with discipline and conscientiousness.

Wild, resource-rich, and huge beyond imagination, Alaska now pulled at me. Away from fishing and the sea, and far from the ports of Alaska, I found myself missing the electric moments of expectation as each new crab pot broke the surface and offered up the luck of the catch. I missed working in a land as breathtaking and unaffected as any on earth and the "go to hell" spirit of that unrestrained life.

On the road, in church, or fishing near Mt. St. Helens lake with my sister's little boy, I could recall it all—the puff and wheeze of the exhausted humpback whale, the screeching cackle of the bald eagle, the clack of grizzly bear teeth chomping through the thick red meat of a salmon's back, the thunderous explosion of a storm wave collapsing across our iron-bark deck, and the excited hoots of our crew as a pot plugged with king crab rose clear of the rich green ocean and swung aboard.

I missed the rush of excitement. I missed the fright. I missed the early-morning clamor of boots, the whining hiss of rain gear, the mild gong ring of feet clamoring up the metal wheelhouse stairs, the metallic clip of the watertight Navy door slamming closed, and the unpredictable moods of the sea itself.

But the most recurring memory of the north was not of money but of the cathartic release of interminable work. It was that satisfying moment at the day's tired end when sleep swept over you in soothing and ever-deepening waves. For as you drifted away into that soothing stupor of exhaustion, you knew that for the moment your part had been done and that some dangerous thing once again had been survived.

As you fell off to sleep, you could sense the accomplishment, like one who had transcended the common bounds of labor. And at that moment, you felt that some significant goal had been realized, some challenge met, some dangerous gamble undertaken and survived, whereupon, lying in the tossed darkness of your bunk, a mere breath from sleep, your sense of relief was almost ethereal. And you could whisper secretly to yourself, "I did it!"

179

TWELVE

When the 1980 winter tanner crab season got under way in early January, the crew of the *Rondys* once again gathered in Kodiak, ready to do battle. We'd hardly begun work fishing out in the turbulent Kodiak Island waters of Shelikof Strait when a bitter cold front moved in.

During the day on deck, we squinted into bitter arctic winds that knifed through our gear in gusts up to thirty-five knots. Carrying a chill factor of minus twenty degrees Fahrenheit, the penetrating cold forced us to keep moving. In an effort to combat the numbing effects of hypothermia, we wrestled with one another on deck over the right to perform certain jobs. During the night, we pointed the *Rondys* into the husky sharp-crested waves of Shelikof Strait, and working in two-hour shifts, we watched for the unwelcome appearance of ice floes. Scores of the jagged white chunks, some several miles long, had reportedly broken free up in the Kamishak Bay-Cook Inlet country and were drifting toward us in the swift Alaskan tides like nature's own battering rams.

Here, on the west side of Kodiak Island, those slabs of frozen sea had already carried away dozens of crab pots and had sliced through the hulls of several crab boats. From then on, especially during the long January nights, eluding the lumbering wedges of ice became a skipper's highest priority. While the fishing had proven skimpy for most fishermen, in only five days of moving gear, our skipper tracked the tanner crab as methodically as a man might follow footprints in wet cement. And our catches increased from ten or fifteen "bugs" (crabs) per pot, to where they were producing over one hundred per pot. For several days, our pots rose with impressive waist-deep piles of the brown and lanky tanner crab washed into her bottom end. In spite of the mean elements of weather and spray-ice forming on deck, we moved with excitement and vigor and without complaint. When we had earned $750 a day per man for four consecutive days, we concurred on deck that it looked like the kind of season most other crews only dream about. Barring some mechanical catastrophe, there was no reason we couldn't pocket twenty thousand dollars apiece in the next few weeks.

Then on the fourth day, in the midst of our fine streak of fishing, Vern's calm but disappointed voice sounded unexpectedly over the loudspeakers. "Pack her up, boys. Pull in the block and secure the deck for running."

As deckhands, we were used to an occasional prank.

Then he added, "The Bering Sea ice pack is moving south. It's already pushed down into our gear. We've got to go save what we can. The *Progress*, the *Juno*, the *Golden Viking*, and a whole passel of other boats are already headed that way."

In stunned silence, we rushed to secure the deck. Minutes later, the *Rondys* swung around and we were headed west on another seven-hundred-mile come-what-may adventure to the Aleutian Islands and points north.

At the end of the 1979 king crab season, we'd left approximately 240 of our large "seven-and-a-half-bys" with their lines shortened, bait removed, and trap doors tied open in what was known as a wet-storage area several hours northwest of Port Moller out in the Bering Sea. Now, giant polar ice floes, some tens of miles in diameter, had drifted abnormally far to the south from the polar ice cap and into the fleet's designated pot-storage area, wiping out crab pots by the thousands.

We crossed the waters of Shelikof Strait and ran with a rough gray sea. Over the next two days, we passed by the mouths of Puale Bay and Wide Bay, nature's own wind tunnels, and as always, the icy wind sharpened briskly. Six years before, during the same time of winter, knifing cold williwaw winds accelerated through the eight-thousand-foot mountain passes and came roaring out the mouth of Puale Bay, creating some of the worst icing conditions to be found on the face of the earth.

It was here during the howling winter of 1974 that the famous *John and Olaf* tragedy took place. At the time, the bitter williwaw winds were accelerating to more than 120-knots as a wintry Bering Sea pressure system came funneling through the high mountain passes between the Bering Sea and the Gulf of Alaska. Packing a chill factor for which there are no charts, the blasts of wind moved with sustained speeds that would compare with many of history's worst Gulf of Mexico hurricanes.

George Johnson—better known as "Turkey George"—was skippering the 170-foot *Shelikof Strait* when the U.S. Coast Guard called and asked him for his position. "I was fishing tanner crab in Kukak Bay," he recalls. The Coast Guard wanted him to proceed to the *John and Olaf* immediately!

"Well, first," shot back Turkey George, "you're going to have to tell me exactly where the *John and Olaf* is!"

When he finally turned the corner into Puale Bay, he found himself driving into what he described as a "blinding, freezing ice fog that was blowing one hundred and fifty knots, anyway! I don't care what anyone says!" he recalls. "I was there and I'm telling you I've never seen it blow like that. It was throwing solid water at us, and it was freezing as fast as it hit. A total whiteout!"

The williwaw winds wasted no time in blowing George and his ship backward out of the bay the same way they'd come.

At about the same time, U.S. Coast Guard C-130s were attempting to fly sorties above the Maydaying shrimp boat *John and Olaf*. At no less than a thousand feet, they encountered freezing sea spray and were forced to turn back.

For two days, Turkey George tried to get in close enough to help, but it was no use. He could only get so far and then the wind would throw his huge steel bow one way or the other. Once it got turned and the driving weight of the hurricane-force winds came full against the entire broadside length of the ship, it blew the huge crab boat backward as quickly and easily as if it was drifting out of gear.

Such a physical reaction was understandable in light of the fact that even a 120-mph wind would exert a force of approximately 112 pounds for each and every square foot of exposed surface. Calculating the total wind pressure against Turkey George's ship under such wind conditions (10 feet in height by 170 feet in length by 112 pounds per square foot) suggests that each time his bow swung around, turning him broadside in the water, he was struck by a wind force of approximately 190,000 pounds!

It was during the second day of his efforts that Turkey George opened the wheelhouse door and stepped out onto the tiny deck of the crow's nest to urinate. Suddenly, as if struck by a car, the typhoon-force winds blew him against the back railing and nearly knocked him unconscious.

Staggered by the blow and numbed instantly by the unbearable cold of the arctic winds, Turkey George reeled back toward the wheelhouse, to find that the door he'd exited only seconds before had already frozen solidly shut. None of his crew was aware of his predicament, he knew. They were below in the galley eating at the time. If he waited for someone to meander up into the wheelhouse and notice him missing, he knew he would freeze to death.

Bracing himself against the gusting squalls, he climbed slowly down over the outside of the railing. Moments later, he shocked everyone seated in the galley when, shivering uncontrollably, he came racing in off the deck and collapsed in front of the wall heater.

It took forty-eight hours for the Coast Guard cutter *Citrus* to battle through the elements and reach the *John and Olaf*. Stranded on a reef and encased in tons of ice, she was a haunting figure. She looked more like an iceberg than a shrimp boat. The ship's life raft was gone, and the men, too. A cup of coffee sat undisturbed on the counter.

For those on board the *John and Olaf*, the decision to abandon ship and set out across the wind-ravaged sea in a rubber life raft had meant certain death.

Though the remains of the life raft would be found several hundred miles away on the barren shores of Chirikof Island, none of the crewmen were ever found.

As the *Rondys* powered ahead, twice a day, at 8:00 A.M. and 6:00 P.M., Peggy Dison (WBH-29 PEGGY) reported the weather for the National Weather Service. Afterward, she sent and received messages. "All mariners! All mariners! This is WBH-twenty-nine with the Alaska marine weather," she would begin, broadcasting from her office in Kodiak in her familiar voice. Then came the date and the broadcast information.

"It's one of the few times when you won't hear the voice of some Norwegian chattering away on the radio," joked Sampson as we passed Sutwik Island.

By now, we had all come to know Peggy, at least her voice. Later, I would visit her back in Kodiak. The wife of legendary skipper Oscar Dyson, she was an intelligent and gracious host.

Over the years, WBH-29 PEGGY had become as much a part of the fleet as any of us. She'd paid electrical bills for worried fishermen too far from home to beat the electrical companies' cutoff date. She'd ordered spare parts for numerous vessels and arranged plane flights to carry the parts out to the remote bays where boats were stranded. She'd delivered the good news of newborn babies, passed on to the Coast Guard the latest coordinates of fishing boats signaling for help, calmed skippers too terrified to share useful information with the Coast Guard, and wept at fishermen's funerals.

It was Sampson, our quiet, steady deck boss, who said it most succinctly: "She's one of us."

Night and day, we moved along at a ten-knot pace, following the untouched sweep of the rugged, snow-covered terrain of the Alaska Peninsula. It was hard to recall that in summer these same white lowlands now encased in ice and snow and chilled by incessant arctic winds had been bathed in a lush emerald green growth of vegetation, almost tropical in thickness. For a thousand miles—from Anchorage to Dutch Harbor—lay a vast stretch of untouched wilderness sculpted by nature into magnificent forms of disintegration.

As we neared False Pass, we learned that for the first time in decades it had frozen over. So we steered clear of the shortcut and began the eighteen-hour journey around Unimak Island, an island laced with volcanos such as Mt. Pogromni and Mt. Shishaldin.

When we reached the westernmost tip of Unimak Island, we headed into Unimak Pass. From here, the Aleutian Islands stretched westward toward the Soviet Union's Kamchatka Peninsula in a sweeping arch of islands more than a thousand miles in length. Next, we passed Cape Sarichef, a famous point of land that overlooks some of the worst waters on earth. Old-timers sometimes referred to this stretch of water as the "Isle of Lost Ships."

It was at this very spot that on April 1, 1946, the largest tsunami ever to strike the North American continent came ashore. Caused by the sudden shifting of undersea plates, the force of any given tsunami can travel virtually undetected across thousands of miles of open water at speeds approaching five hundred miles per hour. As they near land, they generally slow and steepen, occasionally developing into some of the most colossal wave forms ever seen on earth.

It was just such a wave that came ashore on Cape Sarichef. The five coast guardsmen asleep in the lighthouse more than one hundred feet above the sea probably never knew what hit them as the concrete lighthouse and much of its foundation were torn up and swept away by the killer wave. Rebuilt in 1950, the new lighthouse was anchored in solid rock some 170 feet above the shoreline.

Stories have been passed down about the place and the land's effect on the lighthouse keepers who have been stationed there through the decades. It is said that many of these keepers, left to themselves for months on end, eventually lose touch with reality. Personally, I could not have imagined a more desolate or lonely duty.

A few years before, in this same stretch of water aboard this very ship, Vern Hall was at the helm when he found himself caught up in a strange natural phenomenon. It was about 3:00 A.M. on a foggy morning, and the crew was busy dumping gear when suddenly a half dozen birds flew into the ship's mast lights.

It was a common occurrence. Crab-boat skippers were accustomed to flocks of seabirds regarding the brilliant white beams of light as some heavenly vision beckoning them on.

Soon, however, more than a dozen more birds came skimming their bow and once again crashed in an identical pattern of feathers and crumpled, falling bodies. Seconds later, when another fifty or so birds struck, it seemed that the dead and meaty duck bodies were dropping everywhere.

No sooner had those birds tumbled to the deck than perhaps a thousand more of the misguided birds came pouring in at a good fifty miles per hour. On they came, thousands more crashing suicidally into the boat, the lights, the windows, the wheelhouse walls, and the crewmen themselves; and suddenly, the strange drama began to take on an element of danger.

"Will you look at this?" cried Vern Hall over the deck speaker. Looking up into the mast lights "was like looking up into a blizzard of speeding bird bodies!" he recounted.

At any moment, they expected the maniacal onslaught to subside, yet on they came. In the next few minutes tens of thousands arrived.

"You guys come on inside," ordered Vern Hall.

Upstairs in the wheelhouse, "it sounded like a war going on," recalls Hall. "We could hear them hitting the roof, hitting the windows, the antennae, the deck lights, the walls, everything!" They plowed into the deck, thudded against the wheelhouse windows, and ricocheted off the hand-railings and mast cables. Working on deck, Bill Jacobson remembered the "whizzing sounds of bodies" as they careened off the wheelhouse walls at fifty miles per hour.

The pesky birds found their way down into the engine room's breather pipes and wound themselves inextricably into the fans, hopelessly jamming them. Their feathers clogged the bilge pumps for months.

Then someone made the mistake of opening one of the two side windows "and a couple of them flew right into the wheelhouse!" recalls Jacobson. "And they were noisy things, every one of them. Some kind of mutton bird, I believe."

Jacobson recalls "going downstairs and opening the galley door, just for a quick look-see." Faced in that moment with the strange and tumultuous confusion of a waist-deep gathering of squawking, fighting, fluttering seabirds, he instantly jerked the door closed. In that short a time, a half dozen of the frantic creatures managed to become jammed in the door.

The frantic cacophony was deafening and hysterical. "It was insane," continues Jacobson. There was just no way to describe the thousands upon thousands of seabirds all milling and squawking, flapping and climbing upon one another as they tried frantically to reach the upper ground. "They were three deep across the whole surface of the boat."

Vern turned the mast lights off and flipped on the small deck light. Then they waited. The flood of birds continued for another ten minutes before their numbers began to ebb, at first to a trickle, then subsiding altogether.

When it was safe to venture outside, there were few volunteers. No one wanted to step outside into the shrieking racket of seabirds. When the first of the crewmen reached the deck, they found it slick with droppings, matted with feathers, and covered with tens of thousands of seabirds, alive and dead, crippled and dying. Disoriented and frenzied, the birds scratched, pecked at, flapped against, and defecated on everything within range. It was sad, but there was nothing to be done but try and return them to the sea. Wearing work gloves and rain gear for protection, the crew began shoveling the whole frantic, pecking lot over the side.

Long before we reached the official pot-storage area in the Bering Sea, we found ourselves dodging ice that had drifted south from the ice cap. We soon discovered that crab pots left in the designated storage area had been run over with ice and totally wiped out. We were not alone. As many as ten thousand crab pots (millions of dollars' worth of gear) had been lost and now lay hidden beneath the shifting ice floes.

In the pot yards of Seattle or Newport, Oregon (where the solid steel one-inch crab-pot frames are welded together, rigged with nylon webbing, and fitted with the necessary buoys, lines, bridles, and straps), the cost of a single crab pot represented a minimum investment of five hundred to six hundred dollars, depending upon the rigging used. When one calculated the cost of transporting each pot several thousand miles north from Washington or Oregon, however, that price easily doubled. Worst of all was the real cost, that of the crab they could have caught over the next several years of fishing.

During the next few days, we wandered along mile after mile of the irregular outermost edge of the ice field in search of our pots. We had left two boatloads (approximately 250 pots) stored in the area, and were now faced with the bleak realization that none of our gear remained where we had stored it.

If the loss of one or even a few dozen crab pots could be considered a financial setback to a fisherman, the loss of two hundred, three hundred, four

185

hundred, or more of the critical crab traps was an absolute disaster. It was too painful to even contemplate what they might have caught over the next few seasons.

Our pot buoys, we soon discovered, were pinned beneath the broken and shifting masses of ice. Over the past few days, whenever the floes parted, they would bob back to the surface, where the tough plastic Norwegian-made buoys were systematically crushed, their hard orange hides exploding under the incredible pressure as if they were no more substantial than Bugs Bunny balloons.

While virtually all the air-filled plastic buoys exploded and sank under the tremendous pressures of the ice, a few of the resilient white "sea lion" buoys remained afloat. They were made of Styrofoam, and although sea lions often enjoy using their sharp fangs to pop the air-filled buoys, they can do little damage to the foam balls. In many cases the ice sheared the sea lion buoys in two or broke them into crude pieces. Yet the crab pot's ⅝-inch line often remained imbedded in the buoy's soft foam fabric.

For days, Sampson, Capri, and I stood on the top deck of our bow in the frigid arctic air and scanned the horizon for the tiny specks of white buoys drifting among the ice floes. The pinching cold of an arctic wind rushed across our fully exposed bow deck at twenty knots, stinging us with a chill factor of some minus thirty-five degrees Fahrenheit. The brutal impact of the wind and cold sent involuntary rivulets of tears streaming back into our yellow hoods. Every hour or so, whenever our hands or facial skin became dangerously numb, someone would race inside and whip up a pot of steaming hot cocoa.

Each time one of us would sight a buoy, he would raise an arm up and down and, like a man leading a military charge, point out the pot's position. Vern would maneuver in through the meandering islands of ice and, stiff with cold, we would file down the bow ladder, haul the pot aboard, tie it in place, then return to our places on the bow. Some of the pots trapped by the ice had been dragged over twenty-five miles.

Near the close of the third day, I found myself squinting into a raw and windy January sunset. From my chilly perch, I took in the broad, flat floes of ice rubble stretching toward the horizon, and watched as the colorful late-evening rays of light washed over the drifting white of the ice, our ship, and ourselves, bathing us in a strange Day-Glo world of amber and orange.

Though comforting to the spirit, the pale light of the setting sun carried not so much as a hint of warmth. By squinting into the low horizontal glint of the January sun and sighting far across the ice, I could make out the irregular black lines of the shadows of pressure ridges, where converging floes of ice had collided, ruptured skyward, and then frozen again.

Suddenly, a bout of the uncontrollable shakes came over me. I stomped my feet and flapped my arms against my sides in an unsuccessful attempt to warm myself. In spite of the stupefying cold, the thought of my calling it quits was unacceptable. It would be dark soon and Vern would call off the search for the night.

Then I looked over at Terry. His mustache, eyebrows, and beard were

186

frosted white with breath. Retreating back into the once-warm folds of my hood, I secretly abandoned the search. As if in a vacuum, I heard Terry swear, and turning, I caught the pop of miniature waves slapping against our midships. Then came the brittle crackle of the sea spray as it froze in midair, followed by the tinny rattle of ice pellets scattering across our deck. And, glancing behind me, I took in the teardrop beads of frozen spray lying on our deck, shot through with the pale orange light of winter.

During the nights, we slept heavily as the *Rondys* drifted silently amid the tranquil waters near the pack edge. Daybreak often found us surrounded by broad white islands of ice, a silent and eerie confusion of ice floes and icebergs adrift in various states of disintegration. The search proved long and arduous, the experience well outside anything for which a man could have prepared himself. Yet, as we worked on the northernmost edge of the polar ice pack, I came to know the men beside me better than ever before. They were tireless and good hearted and exceptional; the best crewmen I had ever known. Throughout those interminably cold days, they endured without flinching or complaint. And as we did, we labored almost telepathically, each one of us quietly withdrawing into our hypothermic selves, turning over and over in that tired gray matter all that had driven us there and all that now drove us on.

THIRTEEN

Ten days of effort produced a full deckload of some 120 crab pots, which rose in multiple layers more than twenty feet above the water. When our load was complete, we ran them in closer to shore, stored them, and then headed back for Kodiak, and home.

In Unimak Pass, we encountered a heavy swell and a strange and nasty tide-spawned chop that ran contrary to our path and across our bow. It was understandable, for it was here that the Gulf of Alaska waters pushed up out of a massive trench five miles deep in places and poured into the Bering Sea over an undersea knoll that rose to within a mere 140 feet of the surface.

We passed along the full length of Unimak Island and early the following night spied off our port side the cold blue lights of King Cove bouncing in the distance. The icy ten-degree temperature had not lifted, but the winds moved with a reluctant spirit and as we powered along on small seas rolling in the same direction, the icing proved minimal.

By morning, all that changed, for as we had moved along the shore of the Alaska Peninsula, the winds had risen sharply. As we rolled out from behind the protection of the Shumagin Islands, the seas came around ninety degrees on us and struck us head-on. With nightfall, williwaws of miserably cold air shot down out of the mountain passes and drove into us without mercy. Wind gusts of seventy knots sheared the wave tops from the waves themselves and sent a thick ocean spray flying.

The temperature never rose above five degrees Fahrenheit, and as we crossed in front of Wide Bay and Puale Bay, the wind velocities stiffened sharply, coating the *Rondys* in layer upon layer of ice. The spray blew horizontally through the night and froze fast and hard to every exposed inch of the deck, sides, windows, rigging, and superstructure of our vessel.

It was the kind of blow in which radio antennae grew barrel-thick with ice and snapped suddenly under the load, and in which radar scanners froze in mid-spin. It was under these exact conditions that handrailings and mast cables grew yards thick with ice. Eventually, the danger to our ship and, without exaggeration, to our very lives was that of becoming top-heavy with the tons and tons of ice thickening steadily on our ship. If left unchecked, the weight eventually would capsize us.

As the *Rondys* became encumbered with ever-thickening layers of ice, the huge ship reacted to the excess weight. As she continued to "make ice," she settled heavily into the water, her muscular buoyancy giving way to slow and burdensome movements. Soon, we began to plow bluntly ahead through the waves like an overloaded barge, moving through the drenching, freezing lumps of sea with hardly a waver.

The handrailings, deck, cables, and fo'c'sle cabin were continually awash with the explosions that came over the bow. In the unrelenting winds, icicles of seawater several feet in length grew in sideways slants.

If Vern was worried about our predicament, he never revealed it to us. These were perfect conditions for what the NWS (National Weather Service) calls "very heavy freezing spray." But when, by 3:00 A.M., on that pitch-black night, the ice on deck had become a foot deep, he summoned us.

"You boys have a little work to do" was the way he put the situation to us.

With rough duty ahead, we donned our bright yellow rain gear, stretching it over long johns, multiple sweaters, and wool caps. Last came thick vinyl gloves. Then, armed with sledgehammers, baseball bats, and axes, we stepped out into the elements.

In the hope of heading off a potentially deadly predicament, we swung furiously and without rest, striking out at the hard-packed layers of ice with unbridled effort. As we worked, I could feel fear rising in my chest, and looking around me, I noticed how each of our vaporous breaths exited hard, like the white, jetting swath of a discharging fire extinguisher.

Putting all of my 245 pounds into the swing of a sledgehammer, I found that the thickest of the ice coating the railings gave way suddenly to the pounding or not at all. And where it refused the give, ice specks spit back in my face like steel flak from a woodcutter's wedge.

Over the next five hours, we completed a day's worth of common labor, but it was only the beginning, and throughout the night we interrupted the endless stretches of swinging by pausing to bend and hoist and toss foot-deep chunks of broken deck ice—some a yard wide—up and over the side.

All the next day and into the following night, a cruel wind blew down out of the mountains, turning our beards into frozen deformities of hair and ice. We warmed ourselves with steaming hot cocoa, energized ourselves with "black chowder"—mud-thick cups of coffee—and fueled ourselves by gobbling down "deck steaks"—candy bars.

During the glinting black darkness of that second night, I could see crystals of spray-ice blowing horizontally over us. Then I spied the ever-growing forms of icicles protruding down like fangs from the mast cables, and I took stock of the unreal appearance of their forty-five-degree slant.

The hood of my raincoat encircled my face in a tunnel-like form, except for the tip of my chin and beard, which protruded down from the bottom. The sleet striking my raincoat echoed with the sound of hail pattering the side of a tent. And through the hollow of my hood, I could hear the echo and rhythm of my own labored breathing, the plastic rustle of our rain gear, the crunch of frozen spray underfoot, the curses and grunting efforts of my partners, the deafening gusts of wind, and the wet, clinging splatter of the ocean spray as it hit and hardened on the deck and on me.

With feet widespread and eyes watering continually from the wind and bitter cold, I could feel the deck leaping and rolling underfoot. Whenever I rose above the partial protection of the bulkhead side railing, the weight of the forty- and fifty-knot winds drove heavily into me.

Once, during the ordeal, I paused, hunched over and winded, and noticed Sampson doing the same. Now melting from the heat of his sweating body, the ice that had coated his rain gear slid off his jacket deckward in small islands.

Sampson was in a quiet, numb-faced, self-preserving kind of state. I knew we were both wondering whether there would ever be an end to it. I don't think he was even aware of the four-inch icicles clinging to his thick black beard.

He was kind of a cute and comical figure, and I told him so. He kindly suggested I do something vulgar and impossible to myself.

"Whatever suits you, of course," he added ever so politely.

"We gonna make it, Terry?" I asked, changing the tone and subject.

"You can bet on it, Spike," he answered with a quiet determination.

Just then, a long, thin piece of tubular rigging ice weighing perhaps thirty pounds broke loose far overhead and slammed into the deck between the two of us. The ice struck like a piece of concrete, and countless other fragments rained around us. They scattered across the cleared portion of the deck with a strange and musical rattle, very much like wind chimes.

"Jesus! We gotta keep out from under those cables!" warned Sampson. "Do you know what that would have done to your head?"

Vern was studying his map at the moment and hadn't caught the brief event. We wanted to finish clearing the area beneath the boom cables, and so, knowing full well that he would call us inside after such an incident, we said nothing, posted Dave as a lookout, and hurried on to finish the task.

Years later, Vern would tell me what god-awful fools we'd been. He said, "A boat returned to Kodiak from Dutch Harbor in the dead of winter, and its steel railings were all bent down from all the ice crashing down out of the rigging."

Later that night, we anchored off in a barren, wind-beaten bay that carried a Russian name I could not pronounce. Standing inside the fo'c'sle, I listened to the lovely, deafening, armor-eating roar of chain link slapping against bow metal as three tons of anchor and chain spilled overboard.

190

Some thirty-three hours had passed since the start of our battle with the ice. Tired beyond anything I had known, I hung my rain gear to dry, slipped out of my steel-toed boats, and staggered down the hallway to my room and free-fell into my bunk. I had nothing more to give. I lay back in the close darkness, feeling the soft comfort of the position, sensing the effort of my breaths, and took solace in the warm, quiet safety of my refuge.

It seemed that my eyes had only blinked shut when I felt Sampson shake me awake. Two hours had passed, he claimed, and it was my turn on anchor watch.

"Go to hell," I said fondly as I rolled out and reached for my pants.

Upon entering the wheelhouse moments later, I realized that I was not yet fully awake. Strolling to the side window, I jerked it down and stuck my groggy, sleep-swollen face out into the wind and elements.

The idea was to wake up, not do myself in, but the force of the knifing cold wind drove my head sideways and puffed the torso and ballooned the long sleeves of my shirt like plump sausages. The icy shot of air tore at my flesh and watered my eyes, chilling me from head to foot. Before I could react, a navigational map a yard wide lifted off the map table and flew across the room. With my adrenaline pumping, I leapt to close the sliding window and steadied my stagger with a hand against the wall.

Sampson laughed. "Cold out there, now isn't it? I'd be willing to bet those williwaws are pushing eighty or ninety right now."

He paused to allow me time to gather myself.

"We're not forming much ice in here, but it sure is blowing hard. Quite a show going on out there. You better keep an eye on the radar. If we swing around, we're not that far from shore, and we'll drift up on the beach in no time."

When he was sure that I was fully awake, he headed below to his bunk. "Oh, and get Capri up next, Spike," he added, his head sticking out over the top rung of the wheelhouse stairs. "We're taking two-hour watches tonight."

I checked our position on the radar, scratched the distances and loran readings on a notepad, then climbed into the captain's chair. I glanced at the clock, turned out the map light, and began my watch.

Outside, cruel arctic winds were swooping down from the snow-covered peaks of the Aleutian Range, which rose some nine thousand feet into the darkness above us. It was up there, I knew, on this blustery moonless winter night, that an arctic storm front had gathered and was now forcing its way into the narrow mountain passes. These williwaw winds increased to nearly ninety knots as they were compressed into the steep passes. They descended almost vertically, and when they emerged from the darkness and struck the flat black plate of comparatively warm thirty-nine-degree seawater, a thick layer of ice steam arose.

It was a fascinating and eerie display. Like vapor rising from wet pavement, steam appeared to be literally boiling out of the black bay water. Alone on watch, suspended in my leathery seat some twenty feet above the water, I watched the white vaporous forms rise up on all sides. They sprang up and twisted about with incredible speed, engulfing our vessel in a strange Dantesque world of demon figures and swirling steam. Here and there, I could see tiny wind devils spinning across the bay like miniature tornadoes, while tendrils of steam danced up and around the hulk of the *Rondys* like spirit apparitions, dusting her windows and wheelhouse in thin sheaths of ice.

The next day, the *Atlantico,* owned and skippered by Bill Jacobson (once a crewman aboard the *Rondys*), pulled in off the Shelikof Straits and tied up alongside us. It was good to see new faces. Our crews exchanged videotapes, girly magazines, fishing scuttlebutt, and war stories, while overhead in the wheelhouse, our skippers shared information and planned strategies against the weather.

We learned that Mike King had arrived back in Kodiak on a trip north from Newport, Oregon, skippering the 120-foot fishing vessel *King-n-Wing.* When he pulled into port, the ship was said to have looked like something sculpted inside a freezer, with her wheelhouse, deck, and masts coated with more than a foot of rock-hard ice.

While fishing in these same mainland waters, Mike King would later describe for me what real icing was all about. "My pots," he said, "were so covered with ice, you couldn't really see any of them on deck. They looked like ice cubes. And when I launched them overboard, they were so loaded with ice, they floated!"

Aboard the *Rondys,* the sweat and worry of the past days were quickly forgotten when, at dinner, we chowed down on platters of boiled king crab legs, boiled snow crab, fried scallops, and a twenty-pound platter of deep-fried halibut fillets. Around them sat buckets of hot melted butter and tartar sauce, as well as mounds of baked potatoes, baked cinnamon rolls, bread, corn, milk, and soda pop.

As we crowded near the galley table, I saw my chance. I grabbed hold of an especially large and spiny king crab leg and broke it off of a cluster of other legs. Then I turned it over and made a knife cut down the full length of the smooth white underside. Lifting it like a scroll, I pulled apart the sliced edges of the shell. A solid chunk of steaming king crab meat—perhaps ten ounces of it—plopped down on my plate. I poured a small bowl of hot butter over its entire length and devoured the white, succulent, dripping slab of meat, eating it like a burrito.

FOURTEEN

The greater story, however, was unfolding southwest of us, out in the Gulf of Alaska on a storm-raked expanse of sea. None of us who were there would ever forget the Mayday signal sent out by the 110-foot crab boat *Gemini,* or the morbid silence that followed.

The ship and its crew of five had left Seattle during the first week in January. With a moderate load of fifty-eight crab pots stacked on her back deck, she cut straight across the Gulf of Alaska on a northwest course toward Unimak Pass and the Bering Sea.

Their journey went without incident until, some 220 miles from the Aleutians, they ran into a severe streak of ten-degree-Fahrenheit weather and began making ice. In the forty-knot arctic winds, ocean spray exploded over their bow and froze fast to the tall pile of crab pots sitting on the back deck, as well as to her rigging and superstructure. It clung to the thin nylon strands of pot webbing, freezing on impact. The spray spread steadily, filling and thickening until the pot webbing became plugged with ice and took on a square, dimpled pattern, much like the surface of a waffle.

Around midnight on January 9, the skipper, twenty-four-year-old Roy Early, gathered his crew. If left to build up unchecked, the weight of the ice would soon make the *Gemini* top-heavy. Eventually, she would capsize, trapping those inside in a tomb of steel.

"We're icing badly!" the young skipper began. "So get out the baseball bats and let's start breaking ice."

As they drove into the tall and ever-building seas, ice began to form even on

193

their spinnaker window. (This is a motor-driven round portion of one of the wheelhouse windows, designed to spin at remarkable speeds, generating such velocity that in theory no amount of spray or ice can cling to it. The skipper is thereby guaranteed a circle of clear forward vision even in the worst weather.) "You could see it grow!" recalls Wayne Schueffley (pronounced shoo-flee), a twenty-six-year-old crewman from Seattle, Washington. "Then it'd break off and start forming again . . . even though the motor was hot!"

Almost from the beginning, the crab pots that had been stacked on the back deck and chained there had frozen solid in place. Working almost hysterically in the spray-filled wind (one packing a chill factor of some minus forty degrees Fahrenheit), the crew decided to try and break loose some of the ice-weighted pots and dump them overboard to lighten the load.

One crewman made the mistake of climbing up on top of the twenty-foot-high pile of pots, while another ran the hydraulically powered arm of the deck crane across the top of the stack to try and scrape off some of the ice. Then the ship rolled and the frozen stack of 750-pound crab pots broke loose from the deck like a huge block of ice.

Suddenly, the crewman atop the glistening twenty-foot mountain of slick ice began to slide across the top of the stack and overboard. Screaming and clutching for a handhold, he was able to save himself only at the last moment from the deadly plunge into the icy seas when he managed to thrust an arm into a crevice between two pots.

Without pause, the ice continued to build, and when it had grown to a thickness of more than a foot, the young skipper decided to head directly for Cold Bay. But the closer the *Gemini* got to land, the colder the wind became and the harder it blew as it accelerated through the narrow mountain passes lining the peninsula.

On board the *Gemini,* the leaden layer of ice continued to build, thickening to a depth of several feet in places. As the enormous weight of the frozen stuff above the waterline came to bear on the *Gemini,* the ship's center of gravity rose precariously high and the "sail area"—the surface against which the icy and incessant winds could then push—also grew proportionately. If left unchecked, the ship's stability would soon be completely sacrificed.

All day and through the entire night, the crew of the *Gemini* chopped, chipped, pounded, and hammered at the ever-thickening slabs of ice. They flailed their implements against the growing layers of frozen sea, knowing instinctively that their very lives depended upon it. The next day, crewman Wayne Schueffley could see how inconsequential their efforts had been. The ice was continuing to build far faster than it could ever be shucked.

About a day after the battle had begun, an annoying leak appeared in the ceiling of the wheelhouse and water from the waves exploding constantly over the bow began draining inside.

Once as they rose sluggishly up and over a large wave, the bottom of the wave trough directly ahead seemed just to fall out from under them. "The ship plummeted right down into it." And the full weight of the next wave collapsed upon them.

With the sound of a rifle shot, the spinnaker window exploded backward into the wheelhouse. Instinctively, the deckhand on watch ducked—a move that probably saved his life. When he rose from the flooded safety of the floor and looked behind him, he found chunks of the glass imbedded in the wooden face of the wall.

The explosion drew crewmen from all over the ship. Wayne Schueffley arrived in time to watch his skipper turn the *Gemini* around and begin running with the huge storm waves.

In the ever-building seas, the crew worked to pass the rope end of a bumper buoy through the icy wetness of the gaping window hole. That accomplished, they used a winch to suck the buoy up tightly in place, thereby plugging the windy passage and blocking out the miserable chill of the wind and spray gusting into the wheelhouse.

Almost instantly, the buoy froze in place. To ensure that it remained there, they ran the line attached to the eye of the buoy across the floor of the wheelhouse and tied it off on the base of the port-side chair. It would prove to be a providential act.

As the storm intensified around them, the spray continued to fly and the ice continued to build. When the skipper once again turned the ship around, he pointed her into the tall seas. But this time the boat breached and remained there—with its load of crab pots still chained to its stern—stalled on its starboard side. "It was as if there was something holding it there," recalled Schueffley.

Ten seconds . . . thirty seconds . . . a minute passed as an eternity. Ocean waves began breaking over the stern of the *Gemini*. Several crewmen ran to the port side, as if their weight might help raise the huge, ice-laden, pot-encumbered, 110-foot steel vessel.

Then the skipper made a highly skillful maneuver. Turning into an oncoming wave, he cut the ship's rudder hard to starboard, rode it up and over, and used the driving impact of the next wave to bring the ship upright. One, two, three: It was an impressive performance.

Shortly, Schueffley approached the skipper.

"Hey, Roy! Don't you think it's about time we started talking to somebody . . . just to let them know where we're at and about our problem?"

"Yah," replied the skipper. By his tone, Schueffley gathered that he didn't want to call the Coast Guard (an act seen by many skippers as an admission of failure).

When the VHF call to the Coast Guard was finally made, it wasn't in the form of a Mayday; instead, it was to inform them of the *Gemini*'s position. The skipper told them the *Gemini* had breached. It was 3:35 A.M. and his ship was forming ice heavily. A two-day boat ride to the south of the struggling *Gemini*, the crab boat *Aquarius*, a sister ship to the *Gemini*, skippered by Seattle's Mike Roust, was caught in the same storm.

Roust later recalled the high-risk hours of the voyage and the strange collision of open-sea forces. Though he basked in temperatures well above freezing, he recalled that during the storm he saw the "wind gauge peg clear over to one hundred and twenty knots for some eight hours. And it never moved! There were these incredible southeast winds coming in against a southwest wind and they were banking against one another and making seas that stood up like haystacks! There were waves, I *know* they were well over fifty, even sixty feet tall! And we were running with them and they'd come and break right on top of our wheelhouse!

"I positioned my boat so that I had the southeasterly wind kind of on my stern, but quartering almost broadside to the southwest wind. It was about the only way we could ride. And two or three times, we got up on one of those haystacks. I don't know what else you could call them, because the boat just all of a sudden did something weird. I stopped the engines, and all of us who were gathered there in the wheelhouse ducked, and the boat just kind of turned around every which way, with waves hitting us one side and the other, and simultaneously.

"Most of the time, I gave it just enough throttle to keep some control over the ship; otherwise, I know we would have pitch-pulled [110 feet of steel ship toppling end over end] and been lost. There was no doubt in my mind. If we'd gone too fast, the nose of our ship would have taken a dive and we would have been lost."

After their position was transmitted, Wayne Schueffley and the rest of the ship's crew worked frantically to chip out the life raft from under more than a foot of ice. Up in the wheelhouse, one crewman bent to the task of scooping seawater out the back door with a snow shovel.

"To top it off," recalls Schueffley, "we had radio problems. The microphone had a loose connection and to make contact you had to wiggle the cord just so. A five-dollar part in a million-dollar ship! The second radio didn't work at all."

As the crisis became more pronounced, the skipper ordered that the survival suits not be donned yet. He had them placed in his stateroom, which was attached to the wheelhouse upstairs.

"Jesus! When I get to Dutch Harbor, I'm going to fly straight home," swore Schueffley aloud.

Moments later, the *Gemini* rolled again.

"Mayday! Mayday!" yelled the skipper into the microphone, but sliding across the drenched wheelhouse floor, and fighting to raise the ship, he was unable to send out their specific coordinates and never even had time to identify the boat.

It was now 8:00 A.M.—the crack of dawn for that part of Alaska in January—and the crewmen were scattered throughout the ship. Then one crewman on the back deck spotted wave water pouring in over the side. The boat was breaching again. There would be no other chances. Running inside, he began screaming.

"We're going down! We're going down!"

"Get out! Get out!" shouted the skipper to his crew.

Immediately, the deckhands in the galley went for the wheelhouse stairs. As they did, dishes, pans, foodstuffs, and every loose piece of kitchen equipment crashed from the cupboards. "Then," recalls Schueffley, "the walls became our ceiling and floor." The spiral staircase was lying on its side and now presented a puzzling maze through which each crewman caught in the lower reaches of the boat would have to pass.

As Schueffley worked his way toward the top, he heard his skipper trying once again to send out a Mayday signal.

"Mayday! Mayday! *Gemini! Gemini!*" he yelled.

All along the line between Dutch Harbor and Kodiak, fishermen and coast guardsmen had heard the *Gemini's* Mayday signal. When the Coast Guard tried to return the call, we strained our ears to the radio static.

"F/V *Gemini!* F/V *Gemini!* This is the United States Coast Guard *Com-Sta Kodiak*—the United States Coast Guard *Com-Sta Kodiak*, over!"

There was only silence.

The *Gemini's* last reported position was said to have been approximately 120 miles southeast of Cold Bay. While the Coast Guard quickly dispatched two C-130 SAR planes to the scene, and the 378-foot patrol ship *Cutter* into the area, coast guardsmen on duty at the USCG headquarters in Juncau were hurriedly gathering information on the estimated time and location of the *Gemini's* disappearance. When the data had been gathered, they fed the rates and directional flow of the current, tide, wind, and weather into their sophisticated computer system to calculate the probable drift of any survivors.

Soon after the skipper's Mayday call went out, the *Gemini's* engines died. In the galley, the lights faded into a near-total darkness. At that moment, Wayne Schueffley was stuck halfway through the spiral stairway.

"Get your suits on!" he heard the skipper scream. "We're going down!"

The skipper's stateroom was located upstairs. It was connected to the upper wheelhouse cabin to allow him close and easy access. Two crewmen in the skipper's quarters were frantically pulling on their suits when Schueffley made his way out of the stairwell, through the sharply slanting wheelhouse, and into the skipper's stateroom. Schueffley raced to get his survival suit on, but he had only begun when the skipper called out to him.

"Get me a suit!"

No one seemed to hear but Schueffley. Pulling off his own survival suit, he tossed it to the skipper, who, also without a survival suit, was trying to pull the rudder in line one more time. Grabbing another suit, Schueffley delayed donning it and ran to help. Together, he and the skipper threw their combined weight against the heavy wood-spoked wheel. Yet, even as they pulled, the ship rolled completely onto its side.

Perhaps it was the buoyancy of the tons of ice encasing the blocklike stack of crab pots chained to the back deck that created the hesitation in the ship's roll. Or perhaps it was air trapped inside the hull that allowed the men inside a few moments of opportunity to flee. No one would ever know for sure.

As the *Gemini* lay on her starboard side, Schueffley and his skipper found themselves standing on the starboard wall of the wheelhouse, while the wheelhouse windows in front of them rose vertically some thirty feet overhead. The windows around them were flooded with the deep green color of occan. Farther up, Schueffley could see the white foam of breaking waves. It was like being inside a maritime museum built into a raging surf.

Then he looked above him and spied a small square space of winter-gray light at the end of the mine-shaft-like column of the wheelhouse extending overhead. Snow was falling in through the port-side window some twenty-five feet overhead.

Schueffley leaned against a window and pulled at the fabric of his life-sustaining survival suit as hard and as urgently as his hands could grip. But no sooner had he succeeded in working both legs into it when several windows behind him exploded inward from the tremendous pressure of the seawater. In the ensuing wild flood, his suit was swept from his legs and lost.

The gushing water rose fast in the narrow shaft of the wheelhouse, and as it did the skipper turned to Schueffley and screamed, "Go for it, Wayne! This is our only chance!" Then, clad in nothing but street clothes, Roy Early turned and dove straight down into the rising flood of seawater. He was never seen again.

The rushing seawater swept the survival suit from Schueffley's legs, and he found himself trapped in the boiling swirl of current that rushed between the narrow walls and poured down the stairwell into the galley and staterooms below. Then he spotted the rope line, the one used to secure the buoy in the spinnaker window hole. It rose before him like a rock climber's rope and was tied to the chair next to the port-side wheelhouse window far overhead. Schueffley used it to pull himself free of the seawater that was gushing into the wheelhouse (through the starboard-side window) and pouring down the stairwell.

Bracing himself just above the water's reach, Schueffley was working to retrieve his survival suit when two of his crewmates reappeared. Fully clad in the warm protection of their survival suits, they exited the skipper's stateroom doorway. Struck by the swirling flood of incoming seawater, they fought free, and, one by one, climbed madly for the port-side window some twenty-five feet directly overhead.

They ascended frantically toward the top. When the first fleeing crewman came to the window, he tried to break it out with his fist. Then the second crewman caught up with him and thought to roll it down as it was designed. The second man was already halfway out the window, when Schueffley abandoned his efforts at retrieving his own survival suit and began his ascent. It was a "Batman climb," he recalls. "I scaled the wall, climbing with my feet stretched out on each side of me."

Finally free of the death-trap confinement of the wheelhouse, Schueffley crawled outside and got to his feet. He balanced on the pitching, rolling side of the wheelhouse and stood blinking into the gray light. Giant waves were pummeling the listing *Gemini* and launched spray over her entire length. Schueffley felt elated, but it was short-lived. He was sopping wet and clad from heel to chin in soaking cotton clothing. He knew that cotton, unlike wool, was an open-celled material and did little to trap the body's heat. Being dressed in cotton socks, pants, and a T-shirt, he quickly realized, was little better in such freezing weather than wearing no clothing at all, and he could already feel the fabric begin to stiffen in the arctic winds.

Then, to his dismay, he heard the sound of a man screaming. It seemed to be coming from below him. Stooping down on the gyrating platform and shading his eyes from the early-morning spray and light, he peered down into the darkness of the wheelhouse.

"It was like looking down into a poorly lit well," he said. Then he spotted the crewman. The man was sprawled helplessly on his back and his body was spinning in circles. Caught there by the torrent of incoming seawater, he was held fast by the suction of the current as it raced around him and spilled down into the black hole of the stairwell.

Schueffley looked hurriedly around him. The two other crewmen were headed for the life raft now drifting alongside the *Gemini*. He hesitated. Any second, he knew, the ship would roll the rest of the way over, entombing forever anyone left inside. Besides, he hardly knew the crewman screaming for help. He'd never even worked with the man. It was an agonizing decision.

"Help!" screamed the crewman trapped in the currents below. "Help! Help!" he cried again and again.

Unable to resist the man's pleas, Wayne Schueffley reentered the vertical shaft of the wheelhouse and started back down. Sliding down the rope line as he went, he lowered himself to a point perhaps fifteen feet below the window. Then he grabbed the arm of the struggling, floundering man and pulled him free of the water. Then the two crewmen climbed frantically for daylight.

Schueffley reached the window first and climbed into the bitter-cold reality of the outside world. Again, his first reaction was one of elation. I got out! he thought. I made it!

As he balanced on the pitching, heaving side of the huge ship, Schueffley glanced sharply around him. Visibility was limited to a few hundred feet. A heavy snow was blowing in steep horizontal slants, and as far as he could see, giant waves were passing in and out of a low and impenetrable fog.

At that moment, it struck him that he and his buddies were more than one hundred miles out to sea and about to abandon ship, and that they were probably the only ones who knew it. He thought to himself, I got out all right. But for what?

As the rescued man completed his escape, Schueffley pulled him aside to make a request. With the wind howling over them and the pounding impact of the storm waves driving against the long steel hull of the *Gemini*, Schueffley had to lean close and yell to be heard. "Tim [not his real name], I want you to do one thing for me. If we're in the water and I'm going to die, just hold me on top of your suit so that I don't drown. Try and stick with me. And if we're in the water, just hold me in your arms and let me freeze to death. But don't let go of me. I just don't want to go that way." But the young man refused to listen any longer and fled along the length of the *Gemini* toward the raft.

Schueffley followed in his stocking feet and wet clothing, balancing as he went on the *Gemini*'s slick and bucking hull. As he moved through the blowing snow and fifty-knot winds, he could feel the life-sucking force of the freezing five-degree air threatening to freeze him in his tracks.

Seconds later, the *Gemini* rolled completely over. As it went, Schueffley walked around the hull as though it were a giant barrel. When he turned and looked for the raft and his crewmates again, they had disappeared. For all he knew, they were dead. Perhaps they and the raft had only drifted out of sight behind the boat. He had no way of knowing for sure. He was alone.

Then he spotted a small aluminum skiff floating beside the overturned hull of the *Gemini*. Schueffley did not hesitate. He slid down the hull, jumped in, cut the skiff loose, and pushed off. But the skiff had holes in it and as he drifted away, the small craft began to sink.

Suddenly, one of his crewmates came swimming past him. The man was wearing a survival suit but had forgotten to pull the hood over his head. A tightly fitting hood and a snug chin-strap were essential if the man's survival suit was to keep the deadly cold seawater out and protect the man wearing it.

"He was obviously terrified," Schueffley recalls. "He was lying on his back and paddling as fast as his arms could stroke." As he passed by the swamping skiff, Schueffley could hear incoherent screams mixed with the sounds of the wind and slapping water. As he watched, the man paddled off over the heaving seas and disappeared into the fog and snow.

There was little time to reflect, however, as a wave broke over the skiff, flooding and flipping it. Schueffley was dumped overboard and, quickly swimming out from under it, he climbed atop the tiny overturned hull and held on.

He knew that total immersion in such waters would quickly leave him unconscious. "I've got to keep my heart out of the freeze," he calculated as he knelt on top of the inverted raft. Lying there, completely exposed to the wind and weather, he could feel the frightful cold bearing down on him. He noticed that his dripping socks had frozen solid around his feet, and he began to pray.

Moments later, he saw a bright orange object drift out from behind the overturned hall of the *Gemini*. It was the ship's life raft! Designed to protect those huddled inside from the wind and spray and ocean waves, the raft had a bonnet built over the top of it. Such a raft was said to be impossible to overturn.

Schueffley spotted what looked like two figures inside. He screamed to be heard. "Hey! Help! Get your ass over here!"

He could tell those in the raft had heard him, or at least seen him. They tried their best to paddle toward him, but the wind and seas were driving them backward. Helpless to slow their departure, those inside were drifting steadily away. Then he saw one of the crewmen on board the orange raft throw out a coil of line, and soon one hundred feet of it was trailing out behind the raft.

Knowing that to be left behind meant certain death, Schueffley slid off the overturned skiff and swam as hard and as fast as his arms could propel him. Up and over and down the swells, he stroked through the liquid ice until he finally caught hold of the trailing line. He clutched the line with a death grip and felt himself being dragged through the open sea by the two crewmen inside the raft. Finally, he was pulled aboard through the tunnel end of the raft's bonnet.

Inside the raft, Schueffley could feel the miserable chilling cold cut through his wet cotton and into his flesh. From the knees down, the feeling in his legs was already gone. "Pull the flaps closed!" he screamed. "And tie them shut!"

The other two crewmen objected. They wanted to keep a watch out for their skipper and the missing deckhand, so they drifted—and waited. The seawater that had found its way into the raft washed constantly over Schueffley's lower extremities. Shortly, he discovered that his pants legs had grown stiff, rigid with ice. Suddenly, one of the deckhands panicked.

"You got a knife?" he screamed. "Who's got a knife? We're still tied to the boat!"

Schueffley handed him his knife, and the crewman cut the line leading back to the sinking *Gemini*. Immediately, the raft accelerated out of the area and disappeared into the fog. It wasn't until later that they realized they'd cut the line leading to their sea anchor, an important stabilizing part of the raft.

For the first few hours, as Schueffley huddled and shook in his cotton clothes, the others checked for a sign of the missing men. Then one of his crewmates noticed Schueffley's uncontrollable shaking and unzipped his own survival suit, pulled off his sweater, and handed it to him, dripping wet. It, too, was made of cotton.

As the hours wore on, the survivors in the raft gave up hope of ever seeing their two missing comrades again. So they began playing games to lift one another's spirits. "What we'll do when we get back home" was one of their favorites. They'd drink, they fantasized. They'd make love, and eat all the pizza they could swallow. They'd even go to church.

Physically, Schueffley knew he was a mess. He was beyond cold and couldn't stop his body from shaking. At the rate he was going, he didn't know how long he could hold out against the hypothermia. Without a survival suit, Schueffley soon sank into a personal determination to just hang on.

They found a can of K rations stored in the raft. When they tried to open it, however, they managed to break their only knife's blade.

Beyond that, they had reason to worry. What chance did such a tiny raft have against the same bitter wind and monstrous seas that destroyed their crab boat? Inside, on the floor of the raft, the crew could do little but huddle together and pray. The storm winds roared continually across the raft's hood cover and the crewmen were forced to yell to make themselves heard.

Worst of all, after every seven or so hours of dark gray daylight, a total darkness would close over them. During the day, "there was always a little light glowing down through the roof of the raft, but that faded out at just about dusk, and after that you couldn't see your hand in front of your face," Schueffley remembered. As he strained to see in the black nothingness, he couldn't tell whether his eyes were open or closed.

Schueffley suffered from the numbing cold. For a time, he tried to lie on top of the bag of K rations to keep his legs out of the frosty water, but it helped only marginally; the thigh-deep ice water on the floor of the raft continued to slosh over him.

201

Now and then, he found himself studying his numbed legs as though they were objects detached. They had become encased in ice and lay motionless and swollen and discolored before him, and they carried not a trace of feeling all the way to his hips. Then it dawned on him: In the bitter cold, the blood of his legs had finally congealed. The flesh of his legs had frozen through. His legs had frozen—literally!

Convinced that his time was short, Schueffley turned to his fellow crewmates. "Look, you guys, if you don't do something for me, there's no way I'm going to make it. I'm going to die tonight."

His buddies moved quickly to help. One crewman slid beneath him while the other sprawled out across his front. It was a move designed to sandwich the severely hypothermic Schueffley, and throughout the night his crewmates kept up the vigil. Schueffley could feel the cold and wet rubber of their survival suits pressing in on him. But their efforts produced only a slight warming effect.

Despite their efforts, Schueffley grew more hypothermic, his thinking becoming unclear. With the dawn of the third day, he found he could no longer hold down even water and, slumped on his side in the icy bottom wash of the raft, he lay retching off and on all day.

"Hey," he shouted. "Let's set off one of these flares. We've got to be in a shipping lane. There're bound to be other people out here. They'll see us."

It took only seconds for his two crewmates to vote down the idea.

Then they heard a plane. It seemed to be directly overhead. "A plane! A plane!" Schueffley shouted. The silver speck of an aircraft was moving across a break in the thick cloud cover. But the weather was terrible, the visibility poor, and by the time they got a flare out and struck and lit it, the plane was gone.

The aircraft they'd spotted was a USCG C-130 SAR plane making a routine grid flight over the area. Flying blindly ahead through low-lying layers of cloud, fog, and snow squalls, the captain and crew had seen nothing. The pass had taken only seconds.

Now as they squatted inside the small inflatable raft, the survivors of the *Gemini* felt a discouragement that cut deeper than words. Schueffley said, "To hear it, then to *see* it, and for them not to see us was almost too much." The dispirited crew slumped back in disgust.

Hour after hour, they drifted over the shifting gray mountains of water, and the weather steadily worsened. Soon, they found themselves buffeted by deafening gusts of wind, and spray began to tear at the bonnetlike ceiling of their raft.

Already precipitous, the storm waves grew higher, their troughs deeper, their crests more accented. As if to add to the feelings of utter hopelessness and despair, the broken structure of the clouds overhead closed ranks. The snow squalls and deep banks of fog that boiled in over them now hid them from any hope of discovery or rescue. For several hundred miles in any direction, the visibility had diminished to nothing.

Then the dull gray daylight gave way to darkness and the interminable sentence of the third night began. Imprisoned inside the raft, lost in the leaping, roaring night, caught in a tar-black world without hue, hope, or dimension, the survivors held on.

The waves built fast and struck hard. There was nothing to be done but ride them out. Then came a thunderous roar. It was a freak wave, and it was breaking their way. The sound of it terrified everyone. It drowned out the slapping roar of the wind-raked raft cover and intensified dramatically as it approached.

"You couldn't do anything because you couldn't see anything," recalls Schueffley. They could feel it, though, as the crushing weight of the rogue wave collapsed down upon the raft.

Those inside the distorted raft were tossed in a rolling nightmare of bodies and darkness, tumbling together down the clifflike face of the breaking wave. When the unpredictably huge wave finally passed on into the night, all three crewmen found themselves standing on the canopy of their raft, chest-deep in the icy sea.

Advertised as unflippable, the raft remained overturned. The arrival of each new wave sent a charge of water exploding through the canopy opening. They knew someone had to climb outside and try to upright the raft. Even the two crewmen clad in survival suits knew they'd soon perish unless they could right the raft and get clear of the icy Gulf of Alaska sea.

Schueffley knew he couldn't make the trip outside. With his frozen legs unable to support him, he was fighting just to keep his head above the murderous seas exploding through the tunnel-like chamber of the overturned bonnet. When one of the two remaining crewmen refused to go, it placed the responsibility for the journey squarely on the only man remaining.

It would be his lot to crawl outside the raft. In the complete and utter darkness of the night, he would carry out his difficult task against the threat of another deadly rogue wave. To be caught outside the raft by such a collapsing wave would virtually guarantee his being swept from the raft and lost forever to the storm and the sea.

Swimming outside, the brave young deckhand surfaced on the downwind side of the raft. He felt frantically with the glove-covered paws of his hands in an effort to locate the strap built into the bottom of the raft. Placing his feet against the submerged portion of the raft, he screamed instructions to his two companions inside. Amid the roar and crash of the wind and seas, the three of them managed to shift their weight as one and turn the raft right side up.

They bailed water all night and at daybreak discovered that nearly half their K rations and, more important, their few cans of water had been washed overboard.

Back aboard the *Rondys,* we speculated about the *Gemini'*s fate. Covered with leaden sheets of ice, the *Gemini* would find it harder and harder to recover. We reasoned that to regain a great deal of their stability, they had only to jettison the stack of crab pots on their back deck. Perhaps they had done that. Perhaps they had sprung a leak, perhaps the storm was just too much for them, or a rogue wave had broken into their wheelhouse and shorted out their electrical system and radios.

When the third day rolled around and still there was no sighting of the *Gemini* crew, we agreed at dinner that they were probably gone for good. In such weather, even if they did manage to get into a raft and get it launched, they were history if they hadn't been found by now.

Sometime during the night, Wayne Schueffley grew delirious from the cold and passed out. As dawn of the fourth day broke, he awoke and then once again fainted. When he regained consciousness, he discovered that he was lying on the bottom of the raft with his head in the water.

He pushed himself upright then and took stock. What he saw unsettled him. The frozen flesh of his legs looked swollen and white in the ice water. There were large blue blotches scattered across them. They looked so *huge!* His ankles had swollen to about triple their normal size. As for his legs, he'd already given them up for lost, but life would still be worth living without them, he reasoned. Lying back then, he stared at the condensation of their breath, which had frozen into ice crystals on the ceiling of the tent overhead.

From the moment Schueffley had leapt overboard and struck out for the raft, the cold January sea had lapped over him. At that moment, the capillaries in his extremities had begun to constrict and the flow of his blood had begun to slow. The insidious chill of hypothermia had crept over him: fingers, toes, hands and feet, arms and legs, in that very order. On its steady path inward toward the core—the heart, lungs, and liver—it would continue until it rendered his limbs useless. Eventually, the blood flowing through the veins of his body would take on the viscosity of ten-weight crude. Perhaps it already had. Perhaps the heart muscles he'd developed during his years as a marathon runner were prolonging that end.

Though he had managed to keep his upper torso relatively clear of the deadly heat drain of the water, he had no choice but to sit waist-deep in it, night and day. Schueffley knew he was freezing to death. Yet as he lay there, he refused to accept it. Acceptance was the real danger. It was so seductive . . . and deadly. It would begin with an admission of giving up, and once that was done, he knew he would soon die.

"Freezing to death wasn't much different from going to sleep and not waking up," he concluded. Then he remembered his parents. They would never know what had happened to him. The thought weighed heavily on Schueffley and made the idea of giving up all the harder. Then he recalled that his sister was getting married in February. His death would surely put a damper on the spirit of the occasion.

During the days, the men drifted through the rocking, lapping seas in near-complete silence. While the others lost themselves in their own thoughts, Schueffley dreamed. Perhaps they were hallucinations. In either case, they moved over him with such vividness that they seemed real. His first dream involved the torturous vision of orange juice. "I was sitting in a warm and friendly tavern," he recalls, "and they kept bringing me pitchers of freshly squeezed orange juice. But I couldn't have any." He said, "My throat felt parched, so dehydrated."

Toward nightfall, Schueffley began to lose all sense of reason. Lifting a corner of the raft's tunnel flap, he peered out into a gray and blustery day. What he saw flabbergasted him. It was unbelievable! There was a ship floating in plain view for all to see. It looked like a nineteenth-century sailing vessel.

A few days earlier while traveling north aboard the *Gemini*, he'd been reading Jack London's classic novel *The Sea Wolf*. It was a wild and adventurous nineteenth-century story about fur seal hunting up near the Pribilof Islands in the Bering Sea. The ship was skippered by Wolf Larsen, the brutish one-legged existential captain who had passed through these same Alaskan waters more than a century before.

As impossible as it seemed, there, floating before him, was the stout wooden vessel with her cloth sails billowed tight with the full Alaskan winds. "And believe it or not," recalls Schueffley, "there was Wolf Larsen." He was standing up on the ship's deck near her bow railing. And he looked down on Schueffley and said, "You can come aboard, lad, but you'll have to pass through that life first."

Then the haunting poltergeist vision of his ship, the *Ghost*, sailed past Schueffley and the raft, circling once, twice. Round and round they maneuvered. "I really saw him!" he insists.

On the evening of the fourth day as Schueffley lay tenaciously clinging to life, he had his second full-blown hallucination. He came to believe that their life raft had somehow drifted down the entire length of Alaska's Inside Passage, leaving him floating beside Ballard's Fisherman's Terminal on the Seattle waterfront some eighteen hundred miles to the south.

"There were two doctors in white coats standing on the edge of the wooden dock," he recalls, "and they were holding out two steamy hot cups of coffee." It was hard for a guy to figure, but even dockside there in the harbor, the raft was being buffeted about in unusually heavy seas. Sometimes Schueffley could see the doctors and sometimes he could see only water. It's incredible, he reasoned in his foggy mind, we must have floated all the way to Seattle!

In his delirium, he reasoned that with help so near, there was no sense in wasting time there in the cold shell of the raft. Survival suit or none, Schueffley decided to abandon ship and swim to shore. Headed for certain death in the open seas, he had already managed to crawl halfway out the end of the raft when one of his two sleeping crewmates awoke and pulled him back inside.

Wayne Schueffley feared that the darkness of the fourth night might never end. "That eternal darkness, that was the disheartening time." Part of it was grounded in the knowledge that during the daylight hours there was a slim chance of them being spotted. With nightfall, they knew there was no chance at all. It was then that a sense of complete hopelessness descended upon the three remaining survivors.

God, thought Schueffley, they're not looking for us anymore, if they're looking for us at all after so long a time. Then he prayed. "God, oh, God! Let me live! Give me another chance. I'll turn my life around. I'll live for you. Please. Just give me another chance."

Sprawled in the icy wash of the water sloshing back and forth across the bottom of the raft, without the civilized protection of a survival suit or even wool clothing, Wayne Schueffley knew he couldn't survive another night. In a way, he felt relieved. Somehow, it no longer mattered.

That night, phantom images of pitchers of orange juice and imminent rescue flashed through his mind. When the light of morning came again, Schueffley regained consciousness and discovered that he was still holding on. His companions informed him that it was the fifth day, and with it came a sudden break in the weather. The water had calmed to almost flat. When he peered outside, he could see wide stretches of open sky.

The effect on the crew was instantly rejuvenating. For the first time in more than one hundred hours, they were able to gaze upon colors other than the shaded versions of black, gray, or snowy white that had dominated their world. Though it was still quite cold, they decided to roll back one of the canopy flaps and embrace the weather. As they did, weak but nevertheless delightful shafts of winter sunlight filtered in through the small opening.

That afternoon, one of Schueffley's crewmates spotted a freighter far off on the horizon. "It's a boat!" he screamed. He wanted to paddle toward it, but the freighter was barely visible on the horizon and was quartering away from them. It was hopeless.

That afternoon as the slanting winter sun reflected upon the lolling blue of the surrounding waters, a U.S. Coast Guard C-130 SAR plane based out of Kodiak flew through the area. The remnant crew of the *Gemini* had but one flare left when they heard the plane growling toward them from out of the distance.

"Quick! Get the flare! Get the flare!" one crewman screamed as another set it off and held it outside.

The flare spewed out orange smoke, and seconds later the Coast Guard spotted it. Honing in on the drifting raft, the plane began circling continuously. It dropped numerous flares and remained on the scene as it awaited the arrival of the USCG's SAR helicopter team.

Inside the raft, the crewmen were ecstatic. They knew they'd been spotted, and they screamed in celebration. They wept openly and hugged one another, pausing long enough to hand the weak and frozen Schueffley a can of water. They'd been found! Dear God, they were going home!

The freighter they'd seen turned out to be of Korean origin, and at the Coast Guard's urging, it came over to stand by to assist. Schueffley was unable to rise, but he recalls looking out the end of the raft and seeing nothing but a solid wall of steel rising up. It all seemed too much to hope for.

It was nearly dark when the rescue helicopter finally arrived. The helicopter pilot brought his noisy craft down alongside the raft and hovered there only inches above the relative calm of the Gulf of Alaska surface. While his two companions abandoned ship under their own power and scrambled safely aboard the hovering helicopter, Schueffley struggled to pull himself to the end of the life raft nearest the helicopter. When he could go no farther, several coast guardsmen reached inside the raft and physically dragged him from the raft and aboard the helicopter. The second he was aboard, the craft roared into the sky on a beeline course for the airport and base at Cold Bay.

During the flight, the Coast Guard crew encouraged Schueffley to stand up so they could strip off his clothes, but his legs refused to support him even for an instant and he collapsed into their arms. So they stripped him naked and wrapped him in wool blankets and placed a hot steel plate on his chest. In less than an hour, they completed the journey into Cold Bay—the same destination toward which the *Gemini* had been headed when it iced up and rolled over. Upon arrival, Schueffley and his crewmates were transferred to a four-engine C-130 and flown into Anchorage to be cared for by one of the premier frostbite experts in the entire world, Dr. Jimmy Mills.

Wayne Schueffley did not remember the harried flight to Elmendorf Air Force Base in Anchorage, nor did he recall the ambulance ride—except for its conclusion. When the ambulance crew opened up the back doors of their vehicle, a series of camera flashes exploded directly in his face. It was the local press and TV camera crews; he and his fellow crewmates were hot news.

Schueffley lost twenty-seven pounds, one-sixth of his entire body weight, during his ordeal. Yet, inside the hospital, the hardest part was yet to come—the excruciating pain that accompanied the thawing of the frozen flesh of his legs. He screamed for ten consecutive hours.

Medically, the theory went, it was necessary to thaw Schueffley out as fast as possible. They submerged his hips and legs in bathwater as hot as he could stand. Twice, as he lay there half conscious in the hot bath, he heard his own heart monitor alarm go off.

The next day, Dr. Mills told him that his heart had stopped twice during the night. However, he had displayed the most important of all human qualities necessary for survival, mental toughness and a total unwillingness to give up. As one longtime Coast Guard pilot told me, "The will to survive: Some have it . . . some don't."

"Okay, here's the situation," the doctor told Schueffley that afternoon. "You're going to lose your legs, or we can try and use surgery. There's no guarantee that it will work. In fact, we've never done it before, but there's a chance you might save your legs, depending upon the nerve damage that's been done."

For Schueffley, there wasn't any choice. Every four hours, day and night, for a month, nurses wrapped his legs in cloth soaked in a saline solution. When the swelling of the frozen flesh went down, doctors made long incisions along his feet and legs to further relieve the pressure.

Eventually, he was moved to the University of Washington Medical Center in Seattle, where he spent another seven months trying to save his legs and feet from amputation. Schueffley remained in a wheelchair for nearly a year, but through numerous operations he managed to save both legs, although he lost most of his toes and much of the flesh on both feet.

Five years later to the day, inside a snowbound, pipe-frozen eighteen-foot research trailer in north Seattle, Schueffley sat with me and relived the entire experience. For the first time since it had occurred, he recalled the sights, sounds, smells, feelings, and dialogue of the event.

FIFTEEN

Aboard the *Rondys*, where we had marveled at the rescue of the three *Gemini* crewmen, claustrophobic feelings were closing in on me. Adrift in a bleak and lonely snowscape, surrounded by blank stares and faces as pasty white as bargain-brand instant potatoes, the loneliness and lack of privacy had become almost intolerable.

Just that morning, Sampson had come out and tried to make toast. But when the bread popped up in our kitchen cremator, the pieces were slightly overcooked. I watched as he applied the butter to bread as crispy as croutons and black as the shaded side of a charcoal briquette. And what he did next, he did out of spite; I am sure of it. For when he scraped his steel butter knife across the blackened face of the bread, his knife moved with the sound of a rake dragging through gravel. He pretended not to notice, but I knew. And I know he knew I knew. Sampson could be a bastard like that. Then he had the audacity to say "Good morning, there, Spike," as if I was to believe he meant it.

But one couldn't trust Capri, either. A short time later, he sat down next to me at the galley table and began chomping openmouthed through a stack of syrup-soaked hotcakes and yolk-run eggs. The sight was enough to make a man sick. Yet it was his lips that nearly drove me crazy. They sounded like a pair of hands slapping together. And when he chewed, I could see yellow and white chunks of egg sticking to and falling from the roof of his mouth.

My retreat up into the wheelhouse didn't help a hell of a lot. For there I came upon Vern. He was slurping his coffee. It grated on my nerves as effectively as if he was scraping his fingernails across a dry chalkboard. Yes, I had to admit

208

it: Over the past few weeks, I had come to despise just about everything to do with life at sea.

When we rolled into the port of Kodiak again, Sampson, Capri, and I hiked uptown to Solly's for a dinner of deep-fried prawns. We wore boots, threadbare pants, and clean but work-worn sweaters. We sported several months' worth of beard, and our chalk-white faces reflected in the lamplight of the room.

We must have looked like a dubious if not an altogether dangerous lot. Perhaps we smelled badly, too—for as we entered, the room quieted measurably and the clean crowd of land-loving patrons who had been socializing with friends at another table abruptly concluded their conversations and moved out of our way almost magically. They couldn't help but stare, but they were polite about it, parting in a series of clever moves to let us pass, as if that had been their intention all along.

The rough, unbridled lot ahead of me moved like beasts of burden. Hunched forward slightly at the shoulder, they advanced in single file, with their sinewy arms swinging primitively at their sides. It was as if they had only recently learned to walk upright. Then it struck me that these rugged and unkempt creatures were the crewmen off my own ship, and I was one of them.

So I awaited my chance, and eventually found Vern Hall alone in the wheelhouse. He was bent over the map-chart table and writing in a notebook. "Vern," I said, approaching him away from the scrutinizing ears of the others, "I think it's time for me to draw my pay and move on."

Vern rose up and looked at me. "I kind of figured that was coming," he said with a sigh.

"Vern, I've sure enjoyed working for you. And if you can't find a replacement, or if you need me to stay on for another trip, well, I'm game."

Vern tucked his hands into his front pockets, looked down, and shook his head in a casual manner. "No. No, if you think it's time for you to go, then you should go."

Overflowing with nostalgia and unable to put all that I felt into words, I turned and disappeared down the stairwell.

PART
FOUR

THE DEADLIEST SEASON:
FISHING THE GULF OF
ALASKA ABOARD THE
ELUSIVE
AND WATCHING AS
TRAGEDY STRIKES THE
ALASKAN FLEET

ONE

The fall king crab season of 1980 represented the peak of the boom-days explosion. In the shortest season on record, more than 230 crab boats (also a record) now comprising the U.S. fleet caught and delivered the entire quota, an incredible 130 million pounds of king crab in less than a month of white-hot fishing. The fleet's earnings would top $150 million!

That same year, Dutch Harbor became the number-one fishing port in the United States (with $120 million in product delivered dockside), beating out the number-two port, arch-rival San Diego and the huge tuna fleets, while Kodiak was listed as the number-three port in the United States.

Million-dollar earnings and fifty-thousand-per-man crew shares (in twenty-nine days) were commonplace that season and continued throughout the year in a half dozen other supplemental crab fisheries. But the human cost increased dramatically. In the first four months of 1980 alone, fifteen crewmen were lost at sea in Alaska. During the year, more than twenty-eight crab-boat crewmen were killed.

Regardless, with thick wads of cash in hand, a young man quickly got caught up in that greedy tide. The majority of us looked upon the current boom as a laboring man's answer to prayer—one of those freakish moments in history when a working stiff could get ahead for good with nothing more than luck, ambition, and a sweaty brow.

Working on the open deck, exposed to the sea and air, any experienced man knew that there were dozens of ways to die. And during the year, death visited fishermen throughout the fleet without preference or predilection.

The human cost of those times could not be measured merely in body counts or U.S. Coast Guard casualty reports, however, for there was another evil descending upon the land. It came in the form of cocaine.

While hundreds of millions of dollars rolled through both Kodiak and Dutch Harbor, tens of millions of it moved in the form of crewmen's wages. And oh, how the cash did flow. There was something delirious in spending it. A penny-pincher might go all night without buying himself a single drink as newly discovered buddies supplied the booze. The nonstop spending, the waste of it, didn't matter. No man wanted to be looked upon as petty. Besides, a good deckhand could always find work. Everyone knew the king crab boom was going to last forever.

Thousand-dollar nights out on the town weren't at all uncommon. On a slow night, one might ring the barroom bell at $175 a round (for the house), buy a few slips of coke (at $150 a gram) for oneself and friends, make a down-and-out buddy a loan, go on to drink yourself blind, and be in bed by 1:00 A.M., or find yourself stumbling along the seashore in a wide-eyed stupor watching the sun rise.

Go ahead, boys! Drink yourselves into oblivion. Hug that porcelain bowl. Inhale that stepped-on whiff till your nose bleeds and your eyes freeze open. Wrestle and fight until you're too beaten or winded to stand. Lie on the floor and shake your feet in the air. Hate and forgive. Argue and laugh. Race barefoot down Main Street. Run with the dogs and howl at the moon. Relish the moment, boys! For as you have long sensed, deadly serious times lie ahead, and this may be the last bender you ever have, in all of eternity.

Many of those who had come into this fast and "easy" money wore gaudy wide-banded wristwatches with gold nuggets inlaid over every available space. One I saw had intricately designed king crabs molded from gold mounted on it, with diamond-chip eyes staring back from either side of a quartz clock face. Another had diamonds inlaid across its blue oval face in the form of the Big Dipper. Below the Big Dipper, a seine boat had been etched, with a minutely crafted cork line playing out behind it.

Others carried home necklaces, etched ivory, earrings made of whalebone, a superillegal sea lion skin, two hundred pounds of cooked and frozen king crab meat, three-foot-long walrus tusks (found on the beaches near Togiak and still imbedded in the animal's skull), fur-lined cowboy boots, wolverine coats, seal-skin mukluks, and coke spoons and caribou-horn joint clips.

More than a few rushed their cash off to the care of the girl back home— ten, twenty, even thirty thousand dollars and more to be tucked away—to find that when they returned to Seattle, the girl had gone or the romance was over, with every cent spent or missing.

Some literally threw their money away. On one drunken sojourn, three crab fishermen stood on Pan Alaska's cannery dock in Unalaska and, betting who could toss their solid-gold watches (ones with several ounces of gold nuggets inlaid across them) the farthest, took turns heaving them out into the bay.

Back home, some deckhands did invest well and eventually retired. Some bought expensive lots and constructed fine new homes on the hillsides of Seattle or in the San Juan Islands or along the Oregon coast. They bought sporting-goods shops or invested in exotic birds, stamps, stocks and bonds, and rare baseball-card collections.

Scores of others cashed out varying portions of their seasons in buying plush vans, Corvettes, Porsches, Ferraris, Mercedeses, BMWs, speedboats, and a thousand forms of real estate. They bought four-wheel drive trucks, slapped on two thousand dollars' worth of pearlescent paint and lacquer, bolted on turbochargers and all-terrain tires four feet tall, and embraced life away from the sea.

Others took month-long condo vacations (including family and girlfriends) on the lovely beaches and close-cropped golfing greens of Maui or Tahiti or Mazatlán. It was in those off-season months that the once chalk-white fishermen would try to forget the seasickness, the boredom, the sleeplessness, the close calls, and the god-awful loneliness. Lying waist-deep in the azure heaven of some South Pacific lagoon, soaking their dark tanned bodies in a gentle white surf, and swilling down a golden flood of tequila, it didn't take long to forget.

With time, what was even more difficult to remember was that everything you had acquired and all that you desired to hold on to depended on the abundance of an odd, spiny creature wandering the cold and muddy bottom of a wild and desolate sea half a world away.

Yet there were worse ways to spend your money. For too many caught up in the "go to hell" lifestyle back in Kodiak and Dutch Harbor, there was cocaine—grams of it, ounces of it, pounds and kilos of the white and evil powder. And it descended upon the rich, wild, spendthrift crowd of young fishermen like an avalanche.

By 1980, an absolute tidal wave of narcotics had rolled over the major ports of Kodiak and Dutch Harbor. The new generation of fishermen were dumping it on barroom tables, drawing lines with the edges of their driver's licenses, and inhaling it through the strawlike columns of rolled-up one-hundred-dollar bills. The drug represented status and success, and was instantly gratifying, and they used it with the reckless arrogance of youth.

Overnight, it seemed, the drug problem had become larger and involved more money than anyone could have imagined. Secreting it away, they snorted it on boats, in bathrooms, in doorways and abandoned World War II bunkers. The demand and profits were so great and the drugs so plentiful that the task confronting the small local police force proved to be an impossible one.

Once, during the 1979 strike in Dutch Harbor, I had stopped by a boat in an effort to locate a spare part for a hydraulic system we were repairing. Striding into the ship's galley from off the back deck, I came upon a cereal bowl of several ounces (approximately five thousand dollars' worth) of the crystalline stuff, with straws, spoons, and razor blades sitting on a mirror alongside it.

That same season, two young skippers whose combined incomes over the past month-long season had approached $200,000 celebrated their good fortune by chipping in together on an entire kilo of cocaine (approximately thirty thousand dollars' worth). Tying their boats alongside one another, they remained in Unalaska Harbor and partied until their stash had been completely consumed.

Cocaine. Nose whiskey. Sinus candy. The devil's dandruff. Heart-attack

215

powder. Riding the white crest—it was the instant adventure. It was the carefree pleasure that, in many cases, grew into the evil of addiction. With dilated eyes and hearts pounding like jackhammers, fishermen who partook climbed higher and higher. Some claimed that riding the white crest was a thrill more intimate and pleasurable than sex.

While the expensive white stimulant ate away at the septum tissue of the nasal passages, it sent the central nervous system into energetic spasms, and the heart racing, sometimes for days, at 160 beats a minute and more. When the indulger crashed into the valley of dark depression beyond, he often tried to calm the paranoia and the involuntary shakes of the landing with multiple shot glasses of 180-proof Everclear shoved down in quick succession, or six-packs of 100-proof Smirnoff, or 80-proof tequila swallowed hurriedly without the civilized benefits of mixer, salt, or lemon.

Sadly, some crewmen grew to love the drug more than fishing (or the girl back home) and ended up sinking more than fifty thousand dollars (and sometimes far more) into cocaine in a single year.

The use and abundance of cocaine soon became nothing short of obscene. Scores of young men inhaled ounce after ounce of the pocket-emptying powder. In one bar in Kodiak, fishermen sat calmly at the tables and on their bar stools and, turning away to block the bartender's view, used the tiny McDonald's coffee-mixing spoons to sniff the small and delicate piles of powder (a toot to a nostril) up into the far reaches of their nasal cavities. And they did so as easily and skillfully as they might tie a reverse bowline, or whip out a five-second carrick bend knot on the bucking deck of a storm-tossed ship.

It is also important to note, however, that those who did not partake in the use of cocaine (which, I believe, included the vast majority of the skippers and a good many crewmen) often disassociated themselves from those who did and felt a disgust for the waste and pillage the drug reaped.

On deck or in the wheelhouse, the drug was dangerous, as well. A drunk man was a drunk man. A skipper could see it. He could smell it. But with drugs, a skipper just didn't know. Short of urinalysis, there was no definite way to detect it.

When one well-known Bering Sea skipper learned that even the best of his deckhands were dabbling in the risky world of cocaine, he couldn't believe it. They were top hands, tough men. They'd work all day and all night, forty hours straight, and keep right on plugging without a complaint. Nowhere else on earth did men work like this. Yet, he was sure of his information, so he confronted his men outright.

"What's the matter with you crazy bastards!" he began in true skipper form. "You're good workers, but what the hell are you going to do when this crabbing boom is over? And someday it's going to be! And someday you're going to need a loan, and you'll go to a bank and the bank president will ask you what you have for collateral. And all you'll be able to say is that you've got sixty thousand dollars' worth of cocaine stuffed up your nose! What you're doing is snorting away your future!"

TWO

After three years of fishing seasons in which abnormally large populations of harvestable king crab flooded the floor of the Bering Sea, skippers who owned their ships now found themselves awash in multimillion-dollar capital.

Yet even while reaping such exorbitant profits, for most fishermen it would have been impossible to finance the construction of a new $2 or $3 million ship if it was not for the newly instituted Capital Construction Fund program.

The CCF was a federal tax-deferment program in which all the money a fisherman earned could be reinvested without first having to pay taxes. According to the program's guidelines, a fisherman could put off paying taxes indefinitely, simply by reinvesting all that he made. In this way, taxation was deferred until the day the fisherman cashed out what he had built. However, as long as the boat owners kept reinvesting what they made, they paid no taxes at all.

Under this program, and coupled with a quarter of a billion dollars per year in Alaskan crab-fleet profits, ship construction soon exploded. It spawned an entire new generation of huge new high-powered, high-tech fishing boats, which were just beginning to enter the fishery. Motivated by a seemingly endless supply of king crab, unheard-of profits, and incomparable tax benefits, some crab-boat owners seized the opportunity and expanded their private fleet to two, three, six, and even eight boats.

For many, it seemed that nothing could alleviate the hunger they had known through decades of leaner, harder years. Glimpsing the promise of the millions that stood to be made, they ordered bigger and fancier boats built—new $2 million storm-busters, 130 feet long and longer. They were made of the finest

217

steel and came with mural designs of bare-breasted mermaids wearing reassuring Madonna smiles.

These vessels were the new state-of-the-art hunter-killer crab boats. Their sole purpose was to suck up the maximum number of crab off the ocean bottom in the minimum amount of time and with the least amount of effort.

Inside, they added VHS video systems, with movie-library stocks that spanned the gap between *The Sound of Music* and *Deep Throat*. They installed soda machines with all the standard theater-house flavors and stocked their cupboards with packets of microwave popcorn.

In the wheelhouse, they surrounded their plush leather swivel chairs with an assortment of electronic gear that would rival the control panels of a NASA spacecraft. There were multiple assortments of radars, sonars, fathometers, CB and VHF radios, spinnaker windows, engine-temperature gauges, hydraulic valves, and the latest in automatic pilot and loran guidance systems.

Their interiors had opulent color schemes. Some staterooms came gilded in brass and bordered in teakwood, while thick toe-tickling carpets stretched wall to wall. Some had stereo sounds piped into each room, with TV screens so deckhands could watch VHS movies while lying in their bunks, the TV's dial within arm's reach of foot-deep mattresses—mattresses that came wrapped in attractive and genteel-looking cloth.

Kitchens often came complete with counters paneled in hydrocarbons of elaborate color and design. Most sported microwave ovens. Many were equipped with walk-in freezers. Washers and dryers and bathrooms complete with stand-up showers and pulsating massage heads completed the luxury appointments.

Out on deck, they installed automatic coiling machines and herculean cranes with forty-foot arms that could lift and set a 750-pound crab pot into place just so thirty feet away and twenty feet overhead, even in the worst of seas, with no more flexing of deckhand muscle than the push and pull of hydraulic lever handles.

Some of us tried to argue with the times. There was no dignity in being a computer-age fisherman, we claimed. Boats didn't catch crab; fishermen did! But then in the next breath, we were forced to admit that no matter how good or experienced a crab fisherman and his crew were, they couldn't work a fifty-eight-foot wooden limit seiner in the Bering Sea in all weather. It couldn't be done.

With one of the new hunter-killer crabbers loaded with forty thousand gallons of fuel, a hull that could crunch through ice if necessary, and halogen lamps to light up the night, a skipper could run gear night and day, in winds right up to and including seventy-knot and in twenty-foot seas. Working in shifts, he could work his crew twenty-four hours a day, seven days a week, for months on end if he chose—or until his men dropped or mutinied.

By the fall of 1980, dozens of new high-tech crab boats had entered the fishery. And shipyards in Seattle and all up and down the West Coast were working at full capacity to punch out new ones. Each month, newly christened crab boats loaded with three-story-high megaton stacks of crab pots were heading north to cash in on the unbelievable profits to be found in the king and tanner crab fisheries. It was as if no one could remember the days when the quota and the price per pound being paid dockside were but a fraction of the present-day bonanza. And the honeymoon was going to last forever.

Like the vessels they ran, the skippers who captained the ever-swelling ranks of the roughly 230 crab boats within the fleet came in all shapes and sizes and, in this case, states of mind. There were the screamers and schizophrenic kings of mood; there were the benevolent dictators, the autocratic wild men, the manic-depressives, and scores of other acutely ill men. And there were many fine ones, too.

An absolute single-voiced authority aboard a vessel at sea is a necessity. But ask any crewman what it is like to be aboard ship with an abusive or unstable skipper and he'll tell you it is a living hell that can lead to burnout within a few short weeks. Get on the bad side of some skippers and you become the object of scorn and derision for the entire trip. You are the first to take a wheel watch after a long day, the first to be ordered off the deck to prepare lunch or dinner while (it is pointed out for everyone to hear) the "experienced and dependable" members of the crew work on, and the first to be blamed for mechanical failure, dirty dishes, poor food, poor fishing, and bad weather.

Some skippers were masters at the endless psychological prodding of even their best men, thinking that a crewman afraid of losing his job makes a better worker. A few skippers actually encouraged a feeling of conflict and animosity among crewmen on deck.

One deckhand related this anecdote to me.

"So you hate him, do you?" asked one such skipper, grinning. "You hate his guts. You would like to cut him up and toss him in the bait box, eh?" he added, chuckling. "Well, why don't you simply go out there and show him you are the better man? Hey? Work him into the deck! And if that does not settle the matter, you can always go to the fists when you get him back to town." He chuckled again.

Some skippers acquired a near-total indifference to the men who worked for them. "I treat my men like dogs!" one Dutch Harbor skipper bragged to me.

One skipper became so paranoid and distrustful of his crew that each night after work he would lock himself in his stateroom, leaving only to check the engine room or grab a sandwich. Even during the day, I was told, he would lock the wheelhouse doors so that no crewman could enter unless he chose to let him in, and then not without the closest of scrutiny.

Skippers are also paymasters. There were those who paid on time, those that paid late, those that paid early, those who paid only after being threatened, and those who paid not at all. One skipper might hand out several thousand dollars in bonus checks at the end of the year and not utter a word of complaint over the loss of a five-thousand-dollar auxiliary engine but would fly into a frothing rage over the use of too many paper towels.

While a few skippers like Bart Eaton, the skipper and owner of the *Amatuli*, set high standards by paying 6 and 8 percent off the top (meaning that he paid for all expenses including food, bait, and fuel), others calculated the deckhand shares aboard their ships by the use of some strange concoction of greed and voodoo that profited no one but themselves.

Such overbearing or unscrupulous skippers were, however, clearly in the minority. And in the close fraternity of Alaskan crab-boat fishermen, such behavior quickly proved to be a skipper's undoing. Word traveled fast among the fleet and through the ranks of unemployed fishermen ashore, and such skippers and their ships quickly found themselves blacklisted. At any time, experienced deckhands ashore can give you a quick rundown on which berths and vessels to avoid and which to covet.

With hundreds of ambitious youngsters pounding the docks in search of work, and more arriving each day, it soon got to where some skippers were hiring crewmen for one hundred dollars a day, then for seventy-five, and even for fifty dollars for a twenty-four hour workday at sea.

As one longtime Kodiak fisherman put it, "these kids were accidents just waiting to happen." A few boat owners openly exploited their youth and naïveté, and it led to the injury, maiming, and killing of numerous youngsters who found themselves suddenly at sea, thrust into the heat of the action in the most dangerous occupation anywhere, with none of the skills or experience necessary to survive.

On deck, they existed from day to day, terrified youths who fled from each strange new commotion, and often leapt into harm's way.

But crab-boat skippers had their side to tell, too. Be it in Kodiak or in Dutch Harbor, there were vessels that through the attrition of emotion, injury, or incompetence, came suddenly in need of a deckhand. Under these circumstances, hiring a new man off the docks often became a game of "Skipper, beware!"

Accepting a new man at his word, some skippers found themselves a hundred or more miles from port with the third man on their three-man deck too seasick to stand.

One newly hired crewman "retired" after only eight hours of work, claiming that his shift was over. While the others worked on, he withdrew to the comfort of his stateroom cabin, where he showered and made himself a sandwich. Like so many who had come before him, when his ship got back to port, he, too, was forced to pack up his gear and move on. A blindly ambitious young man from Idaho quit his mill job and flew north to Dutch Harbor. With bags in hand, he waited all night on the docks for a chance to beg a berth on board the high-liner crab boat *Amatuli*. When the ship pulled into port, he pleaded with Bart Eaton to hire him, was thrilled when, amazingly, Eaton did, then fell seasick the moment the boat cleared Pilar Rock, and for the entire journey refused to leave his bunk.

The deadly form of naïveté was everywhere. One evening in the B & B Bar, a youngster with attractive blond hair and biceps about the size of a working man's wrist sat down beside me and began pumping me for information. Without so much as ten minutes of experience at sea, he wanted to know how, exactly, a guy could go about getting hired onto one of the better crab boats.

Breaking away from a lip lock on a straight shot of tequila, I asked him what made him think he would feel at home out on the high seas.

"I know this won't sound like much," he replied, drawing nearer as if to confide something important. "But I've surfed down off Huntington Beach [California] ever since I was seven. I feel at home on the water."

A weathered-looking fisherman sitting nearby choked on his drink. I threw mine back and ordered two more.

In an occupation brimming with so many "natural" ways to die, suicide seemed a supremely impatient, if not altogether ironic, alternative. But one such fatal, nightmarish incident did occur aboard one well-known Kodiak crab boat.

How could a skipper have known? The man begged for the job. He looked healthy and seemed to possess a temperate and amiable personality. "Besides, the man said he'd been out before," recalled the skipper. "So I checked around, and, just like he said, he'd gone to sea on several other boats doing fill-in trips. He helped around the boat for a few days while in port, and he looked all right." So, okay, thought the skipper. Let's go.

Then, little more than a day later, as they were nearing the crabbing grounds near the Semidi Islands, things began to come apart. Unable to sleep, the wild-eyed, paranoid new crewman twice had to be disarmed as he wandered the ship packing a twelve-gauge shotgun. The skipper confined the man to his bunk, pointed his ship toward Chignik, and ordered the rest of his crew to watch the crewman.

Several hours later, when things had died down, the crewman crept silently from his stateroom and down the hallway to the kitchen. There he came upon a crewman eating alone at the dining room table, with his back turned toward him. The new crewman wrapped opposite ends of a length of halibut line around each hand, snuck up behind his unsuspecting victim, and, without so much as a word, flipped the loop over the man's head and around his neck and began garroting him then and there. With his wind cut off, the strangling deckhand fell forward onto the table. He couldn't scream or talk. There was no way to call for help.

When the ship's cook came upon the scene, he took in the gruesome color of his friend's face and realized that he was nearly dead. As he raced to help, the berserk new man turned and glared at him, and for the first time, the cook saw how contorted the man's face looked. "It was a demonic expression," he recalled. "It was as if he planned to have me for supper."

Soon, the rest of the crew joined in the struggle. But the superhuman strength of the insane new crewman proved almost too much for even the three of them. Then one crewman yelled, "Let's tie him up!" The wild-eyed crewman bolted. In one quick and maniacal effort, he tossed the skipper and two crewmen off him. They landed halfway across the galley floor. Before anyone could recover, he leapt to his feet and raced through the galley and out onto the stern deck.

Now what am I going to do? the skipper asked himself. He stepped back to defuse the situation, but the new man lifted his legs over the side, sat briefly on the handrailing, then leapt overboard, feetfirst.

"Jesus!" yelled the skipper. "I can't believe it! He jumped overboard! Get the life rings!"

The skipper ran through the galley and up into the wheelhouse. He kicked the rudder over hard to starboard. Then he hurriedly notified the Coast Guard that he'd lost a man overboard.

The first deckhand out on deck spotted the wild man off the stern. The man was treading water and looking up at the sea gulls. He seemed "collected and calm . . . as if that icy water didn't bother him at all!"

The skipper returned to mount a rescue of the man. He would later recall how "he saw me coming around on him, and stuck his head under water about three times, and then just laid there floating in the water. I tell you, he drowned himself right in front of me."

Maneuvering in closer to the drifting man, the skipper ran down onto the back deck, grabbed the grappling hook, and with a single heave managed to hook the unconscious crewman through the belt. Back on board, they tried both mouth-to-mouth resuscitation and heart massage, but without success.

The skipper would never learn what caused the psychic break of the newly hired man. The personal diary the skipper found with his belongings revealed that the man believed he was in the Bering Sea at the time. "Well, I finally made it. I'm in the North Bering Sea." This was the way the skipper remembered the man phrasing it.

The next deckhand hired was a fresh, clean shaven, clean-cut all-American college student from Stanford University.

On a lighter, more humorous note, Rick Williams, the young skipper of the *Alaska Trader,* told me about a crewman aboard a king crab ship that had been working off St. Matthew Island in the northern reaches of the Bering Sea.

The crewman had been on board the crabber for the preceding six months, the skipper recalled, "and during that time he hadn't made a dime." Then he got in an emotional row with the skipper. He had decided he wanted to go home. He'd paddle all the way if he had to. And unfolding a collapsible kayak he'd stowed on board, the young man launched it over the side and set off stroking for parts unknown.

Those on board the crab boat tracked him with binoculars and saw him paddle to the rock shoreline of St. Matthew Island. He set up camp beside a tiny plywood shack built to cover a small diesel engine that pumped fresh water out to the floating processor *All Alaskan.* There the angry youngster sat, refusing all pleas by visiting crewmates to return to the ship.

Shortly, the skipper aboard the fleeing crewman's crab boat radioed the Coast Guard and stated his case.

The Coast Guard's reply? The stranded deckhand was the skipper's responsibility. It was up to the skipper to see that the man got off the island safely. The Coast Guard would not intervene unless the man's life became imperiled.

The fleet wasted no time in dubbing the rebellious young deckhand Robinson Crusoe. Over the next few days, inquiries as to the lad's welfare bounced back and forth across the CB waves.

The youngster's feelings were understandable. He'd worked his heart out for more than six months and hadn't grossed a dime. He'd become disheartened and morose, and those who skiffed in through the fog and choppy seas to plead his return found him immovable. He wanted to go home, and he planned to paddle his kayak the entire way back to Kodiak, if necessary—an outing of approximately one thousand miles.

Finally convinced of the suicidal hopelessness of such an attempt, the deckhand stated that he would not budge from that spot until somebody agreed to return him to civilization.

Four days passed before Mr. Crusoe agreed to leave the beach and return to his ship, but only on the condition that they take him to Nome, where he could catch a plane home. Eventually, both parties agreed. Loading the crewman and his kayak safely on board, the skipper pointed the crab boat east toward Norton Sound.

Along the way, they decided to drop off a load of baited crab pots and do a little prospecting. That done, they hurried into Nome as promised and delivered the destitute, sea-struck sailor on the beach. Then, racing back to the fishing grounds, the skipper of the crab boat paused to pick up the load of prospecting pots he'd dumped on the way in.

When he began pulling the gear, the crew couldn't believe their eyes! Pot after pot rose literally overflowing with king crab! One after another, they surfaced, jam-packed with upward of three hundred king crab each! After six fruitless months at sea, they had struck gold. In the weeks ahead, the crew would pull hundreds of thousands of pounds of the spiny red creatures from the sea. Crew shares would top thirty thousand dollars per man!

Robinson Crusoe, it seems, had given up one effort too soon.

THREE

After vacationing for several months in the South 48, I returned to Kodiak once more and took a job aboard the eighty-six-foot crab boat *Elusive*. After an opelio crab season off the Pribilof Islands in the Bering Sea, I spent a peaceful and contented summer packing iced-down loads of salmon between the fleets fishing in the isolated bays around Kodiak Island and the canneries in the village of Kodiak itself. When the king crab surveys carried out in the Bering Sea revealed that surprisingly few harvestable king crab remained there, I decided to fish the fall crab season in the waters around Kodiak Island.

The day before the opening of the fall king crab season, three other crewmen and I worked in a miserable wind-driven rain to load our crab pots. We craned them aboard, bound them with chains, and stored a ton or so of frozen herring bait in our freezer. Then we packed our foodstuffs aboard and secured our deck for running.

But it was rough weather, and several of the crab boats that had gone before us radioed back that a fifty-knot wind was blowing outside, with sixteen-foot seas pounding in against Chiniak Head. It had been years since I'd seen such seas so close to land, but since our course was to be different from that of much of the rest of the fleet, we left the safety of Kodiak in spite of the weather report.

Our plan was to take the somewhat-protected waters from Kodiak to Ouzinkie and Whale Pass. Yet we had hardly begun the journey when the weather found us. The seas just outside Kodiak's city limits were stacking up higher than any of us on board could recall. And when we passed out from behind the close protection of Near Island, a fast-breaking surf drove into our starboard side, throwing our bow a full twenty degrees off course.

Once committed, there was nothing to be done in the narrow channel. Veer to port and you ran smack into Kodiak Island, only a few hundred feet away. A turn to starboard—and into the waves—also would have taken us on the rocks. A skipper might eventually pound his way around the end of Long Island and out its channel, but that would mean running the entire one-hundred-mile length of Kodiak Island broadside to the high winds and hard-driving seas. Running in the trough with a tall load of king crab pots is one of the most unstable of open-sea maneuvers.

Our radios were mounted on the ceiling overhead, and now they squawked with the chatter of ship captains trying to round Cape Chiniak. I stood with my crewmates in the wheelhouse while our skipper navigated. Ahead of us, we could see the *Kodiak Queen,* with skipper Jack Johnson at the helm. Now as we watched, all 155 feet of her black steel hull lumbered up and over an exceptionally steep wave crest, then disappeared completely into the canyonlike wave trough ahead. She showed herself again as she rose over the steep slope of the next wave. Searing winds stripped draining seawater from her superstructure and blew it into silvery gray veils of mist.

"Oh, I like this!" said MacDonald to no one in particular as we plunged headfirst over the crest of one mountainous wave.

"We could always go back and call it a season," I offered, holding on for dear life.

"Yah, I might just decide to dump this load off in Chiniak Bay and go back and tie up and take you all out for a night on the town, too, but I wouldn't bet on it," replied MacDonald, his voice tightening as another wall of breaking white water closed on us.

The wave struck our starboard with a resounding thud, knocking the vessel sideways for several fathoms. We braced ourselves and held on as a gusty, irregular rhythm of waves broke completely over the ship. Each time, a flood of seawater drained off the ship's superstructure, temporarily blurring our vision.

But our skipper hung fast to his commitment and remained steady to his course until we eventually pounded our way around the last red swing buoy. Then, with the force of the seas directly on our stern, we shot ahead and soon passed through the narrow slot called Ouzinkie Pass. Now we were no longer fighting the wind and seas, and the force of the storm pushed us along on a remarkably level ride. Leaving our skipper to navigate through the narrow waters of Whale Pass, my crewmates and I went below and sacked out in our staterooms.

Now married, the father of four, and a much respected captain in these Alaskan waters, MacDonald was introduced to commercial fishing more than a decade before while working in a management-training program in a large grocery store chain down on the Oregon coast. He "hated the work," he recalls, being indoors all the time, with no hope of significant profit or adventure. Then a friend offered to take him fishing for a trip as a paid deckhand aboard a shrimp boat.

MacDonald was earning about $130 a week when he rode out of Charleston

Harbor near Coos Bay on the Oregon coast for the first time. The shrimp fishing at the time was red-hot and soon Tom MacDonald found himself "making two hundred dollars a *day!* I couldn't believe there was that much money in the whole world!" he recalls. He held his own on deck, and by journey's end his friend offered him a full-time position on board the vessel.

The lure of the money and the sense of freedom and independence of a fisherman's life were too much for MacDonald. When he arrived back in port, he quit the management-training program on the spot. Already, the thought of returning to the regimentation and confinement of such a job seemed incomprehensible. He would never live that kind of life again. And with ever increasing success, he would go on to run, own, or captain such ships as the *Betty A.,* the *Ironic,* the *Sleep Robber,* and now the modern and immaculately kept crab boat *Elusive.*

It was well past midnight when we powered clear of the tip of Raspberry Island. Then, swinging to port, we set a course parallel to the rugged green radar outline of Kodiak Island. And with bruising twelve-foot waves lifting and pushing us on a roller-coaster ride, we sailed down Shelikof Strait.

Tom MacDonald had taken the ship the entire way by himself, and at about 1:00 A.M. he decided to turn in for a few hours rest, leaving me responsible for the ship and crew for a three-hour graveyard shift. Calling it quits, he retired with only a few parting words. "Well, she's all yours . . . Night."

Actually, I had volunteered for the duty. The late-night shift had always been one of my favorite duties aboard ship. It was a quiet time when I could be left alone with my own thoughts and dreams. As I sat alone in the captain's chair in the wintry blackness, my eyes moved habitually between the green face of the radar screen to my left and the seas dead ahead, while a huge sea launched the vessel down the western side of Kodiak Island.

I was deep in my own thoughts when a frantic voice leapt from one of the radio sets mounted overhead. "Mayday! Mayday! Mayday! Help! We're in trouble! Mayday! This is the *City of Seattle!* We're in trouble! Can anybody out there hear me?"

I yanked the microphone from the radio set overhead and was about to call back when another voice cut in.

"Yes, sir, we can hear you! What's your position?"

"Mayday! Mayday! We're going down!" came back the frantic voice.

Then I jumped in.

"What's your damned position? Give us your loran readings!" I yelled.

"We're in trouble! We're going down! Our bow's under water right now!" replied the voice.

Then he seemed to gather himself, and in a trembling voice, he said, "Our position is four-three-five-four-nine-point three and three-two-three-one-three-point five!"

I turned to our map table, and soon I had him pinpointed. He was just two miles due east from us—only minutes away.

As I hurried into the skipper's room, I fought to control the urgency in my voice. "Hey, Tom! We've got a Mayday and he's right here beside us!"

In one motion, Tom rose and leapt to his feet. He paused at the map table long enough for me to show him the distressed vessel's exact position, then he hopped nimbly into the captain's chair and cranked the ship hard to starboard. "Go get the rest of the crew," he ordered.

Then the stirring voice of the frantic fisherman leapt again from the radio set. "We're putting on our survival suits now! We're abandoning ship!"

Ahead of us, concealed by the darkness, drifted the 105-foot crab boat *City of Seattle*. She was reportedly adrift without power or lights. Her interior was aflame, and she was sinking steadily. It would be difficult to locate her.

As we closed on the area, several of us donned our rain gear and ventured out onto our bow. Lying against the bow railings, we searched the chilly black night around us for a sign of the foundering vessel.

In our wheelhouse, Tom MacDonald continued the dialogue with the skipper of the *City of Seattle*.

"You got any flares on board?" he asked.

"Negative," the voice shot back. There was a pause. "Hey you! You with the blue deck crane! You're going right by us now! You're right beside us!"

The fishing vessel *Cougar* was nearby. Perhaps it was the *Cougar* they had spotted.

Then MacDonald switched on the entire array of our ship's huge mast lights and began making frantic circles across the face of the sea, weaving in and out and over and down the passing waves.

Suddenly, out of the corner of my left eye, I caught the movement of smoke streaking across the water, whipped there by the thirty-knot winds. "There it is, Tom!" I yelled, pointing. "Twenty degrees off the port bow! She's burning! See the smoke?"

"By God, you're right!" he answered, resetting our course.

The crew on board the sinking crab boat had been through their own private and unknowable ordeal. It had been a long night for skipper and owner Gary Wiggins. All night, he had idled ahead into storm-generated waves that broke directly over the bow, then washed through the entire deckloads of pots before slamming into the tall steel wheelhouse with a hellacious shudder.

Then, just before midnight, a huge wave struck with a loud bang, knocking the vessel sideways.

The sleeping crew of the *City of Seattle* came alive and pandemonium broke out. Several greenhorn deckhands began running back and forth in the wheelhouse, screaming that they were going to die.

"Shut up! Shut up!" screamed Wiggins. "We're trying to get our position off!"

A crewman was sent forward to check out the bow, and he returned shortly to report that the fo'c'sle and bow deck were covered with some two feet of steadily rising water. The skipper ordered his crew to put on their survival suits.

Wiggins went below to survey the damage. When he opened the engine-room door, he was driven back by a deadly cloud of smoke and Freon gas.

227

Regardless, he rushed below and tried without success to put the generator back on line. But he soon grew too sick to continue.

One crewman held his breath and ran down the stairs into the engine room to see what could be done. When he showed again at the top of the stairs, he was down on his knees, crying from the eye-burning Freon.

When the lights went out, the greenhorn crewmen panicked again. "What are we going to do?" they screamed. "What do we do now?"

"Go topside and get the raft ready!" shot back Wiggins.

"Let's go, honey," said Wiggins's wife. "Let's get off this thing!"

But the main engines were still running, and Gary Wiggins hesitated. "No!" he shot back. "You go! I'm going to stay with her and see if I can save her!"

No one would ever know with certainty what caused the bow to begin to sink, but Gary Wiggins later surmised that the seawater flooding over the bow started pouring into the engine room through electrical conduit pipes that ran the full length of the hull. "When the water rose, it eventually hit the four-hundred-and-forty-volt electrical panel," he told me, "and knocked everything off line—pumps and bilges and lights."

For a time, the engineer could hear the generator running. It sounded as though it was racing out of control. Then the lights and power flickered out altogether, and the *City of Seattle* was cast adrift without power to steer by, or light by which to see or be seen.

Our mast lights threw an icy blue beam of light across the ever-changing face of the cold black sea and the drama unfolding before us. The scene left us silent, awestruck, and transfixed.

The *City of Seattle* was packing a fair-sized load of crab pots. Tied to one another and chained to the deck, the seventy-odd fully rigged crab pots weighed some sixty thousand pounds. Shiny new buoys could be seen gleaming from inside crab pots that had never been fishing. As the forward compartments of the ship's hull continued to flood with seawater, the bow settled noticeably, and ocean waves began to break across her deck. The vessel's tall, steel, stern-mounted wheelhouse was mounted high over the stern. And as the bow slowly sank, the stern rose and the wheelhouse facing out over the long forward deck began to tilt ahead in unison. Well off on our starboard, rolling heavily in the late-night swells, idled the fishing vessel *Cougar*. She was standing by to pick up the survivors.

With smoke billowing from the wheelhouse and seawater rising steadily over the deck, the stern of the *City of Seattle* began to tilt upward sharply, exposing the rudder and propeller.

No sooner had the young crew of the *City of Seattle* finally finished launching their life raft than its sea anchor became tangled in her props. When several greenhorn crewmen once again started to panic, it was Wiggins's wife who cut the line and set them free.

With his wife and crew safely aboard the *Cougar*, skipper Gary Wiggins and one other experienced deckhand finally abandoned ship. One by one, they leapt from the angular tall column of stern steel, their bright red survival suits gleaming in our mast lights.

You could see how the stern was thrust up and down by the waves. The two men fell about twelve feet through the night air and were swallowed briefly by the sea. Beneath the surface, Wiggins heard the *thump-thump-thump* of the diesel engines aboard the waiting ships.

Then, just as suddenly, he was on the surface again. He swam wildly through the jet black sea, fighting against the suck and surge of the *City of Seattle's* stern as it plunged up and down. Wiggins would never forget the fear, the adrenaline, and the long, sweeping valleys of sea. But as he approached the *Cougar,* he noticed that she, too, was heavily loaded with crab pots and was rolling sharply in the steep and irregular seas.

As he drew near, the huge steel hull of the *Cougar* leapt high and free-fell into a trough. She rose, staggered, and fell. Seawater spilled over her bulwarks. The vision terrified Wiggins, and he thought, I don't even want to get close to this boat! But he knew he had to. As he drew closer to the *Cougar,* he could hear the flush and pop of water sucking at the ship's hull.

Then as he maneuvered to within arm's reach of the ship's hull, he saw a big, bearded crewman reach down for him. "All I can remember is that I swam alongside the *Cougar,* and this big bearded guy, and I mean *big,* grabbed me by the back of the neck and with a deep grunt gave me a pull. The next thing I knew, I had landed clear across the deck and was crumpled up in the corner of a crab pot stacked there."

As he lay there panting with exhaustion on the wet wood of the deck, Wiggins could heard the rhythmic flush of water escaping through the ship's scuppers. Then there was a metallic clip as the ship rolled high and the flaps flipped shut.

The large bearded man was none other than Kodiak's Steve Griffing—"Big Steve" to his friends. Standing six eight and weighing 285 pounds, he was exactly the right kind of man for the job. Griffing would soon play a major role in another dramatic rescue attempt that same season.

On board the *Elusive,* we stood by to assist. As the skipper and crew of the *Cougar* maneuvered to pick up the men, our skipper, Tom MacDonald, decided to try and save the *City of Seattle* itself. He disappeared into his stateroom and when he reappeared, he was wearing a diver's wet suit.

MacDonald grabbed a shot of line off the back deck, tucked it under one arm, and climbed out onto the outermost peak of our leaping bow. He sat down and began pulling on his flippers. With the wind filling my ears and spray hissing against my rain gear, I had to yell to be heard.

"Just what in the hell do you think you're going to do, Tom?"

Tom lifted the mask to his mouth and spit on the inside of it to prevent the glass from fogging. Then he turned, pointed to the deck boss sitting at the helm inside the wheelhouse, and yelled, "Tell him to keep this big bastard away from me, or he'll squish me!"

Then, suspended out over the water, Tom balanced himself on the lonely, plummeting tip of our bow. The bow carried him up and up until its momentum peaked and it began its free-fall descent once again.

He adjusted his mask and turned back to me one last time. "Feed the line to me as I go!" he yelled. "I'm going to try to climb up on the bow of that ship. When I do—*if* I can—then tie our two-inch hawser line off on your end and signal me and I'll drag it over to me. Let's see if we can't hook on to that big bastard and drag her into shallow water before she sinks!"

Then as the ship lunged forward into another wave, he clutched his face mask in one hand and the line in the other and fell forward into the sea. MacDonald went under momentarily, came up trailing the ⅝-inch crab-pot line, and set out stroking mightily toward the foundering *City of Seattle*.

While our young crew looked on, our deck boss reversed the throttle and backed the ship slowly away. Then we watched together as our gutsy skipper swam through the undulating sea to the foundering ship. It was a true act of daring, and I did my best to feed him the line free of tangles.

Soon MacDonald swam into a pool of diesel fuel. It was a horrid experience. The fumes filled his sinuses, and the waves washed the oily film over him, coating him repeatedly.

Several hundred feet of ocean away, the huge steel bow of the sinking ship leapt and rolled like a thrashing beast. Up it rose, a full ten feet out of the ocean, dumping untold tons of seawater from its flooded deck. Then it plunged straight down and vanished beneath the waves.

When MacDonald reached the leaping hull of the *City of Seattle,* he paused and backpedaled, as if to evaluate the situation. His body seemed dwarfed by the huge moving wall of steel rising and plunging before him.

For several minutes, MacDonald stood off from the reeling bow, gauging its rhythm. Then he swam closer, knowing that if he was sucked beneath the ship, he would be crushed and that none of us would be able to come to his aid. One could see the calculation in his manner. Then in one motion, he reached out and caught the bow railing. As the bow leapt skyward, MacDonald clung to the outside of the railing and was swept perhaps twelve feet into the night. As I watched, he disappeared into an emerald green flood of seawater draining from the deck and pouring down over him. And then I saw him come under the strain of the rushing current and the burdensome weight of the water, and match them.

As the bow plunged back toward the sea, MacDonald clambered over the railing and clung to the anchor winch and chain as ensuing waves did their best to wash the bow clean of him. Then, during a lull in the wave action, he stood and signaled me. I tied the end of the crab-pot line to the thick two-inch hawser line, then fed it to him as he pulled the expensive high-test hawser line over to himself and tied it off on the *City of Seattle*'s anchor mounted on the point of her bow. When the bow buried itself in another wave, MacDonald floated free of the deck and set out swimming for the *Elusive*.

He was puffing heavily as we pulled him up and over the side. At his adamant instructions, we peeled his wet suit top from his back and stood back as he raced, dripping, through the galley and up the stairs into the wheelhouse, where he threw the engine in gear and began the tow.

He pointed us toward Uganik Bay. Once inside the bay, MacDonald intended to tie the *City of Seattle* alongside the *Elusive* and just run both ships aground. He figured the *City of Seattle,* riding as low in the water as she was, would be the first to strike bottom. When the sinking vessel was resting safely on the beach, he planned to cut her loose and back away. The *City of Seattle* was worth more than a million dollars, and we knew the successful completion of such a task could earn us hundreds of thousands of dollars in salvage fees. But we had no sooner begun our tow when she started shearing from side to side off our stern, veering first one way and then the other. They were long, helter-skelter runs and they made the *Elusive* shudder and shift in the water.

Each time the *City of Seattle* came to the slack end of her run, our thick hawser line stretched elastically, ringing seawater from her expensive fabric like drops from a twisted towel. Then a mile or so off Miner's Point, the bow of the *City of Seattle* nosed toward the bottom and the megaton strength of the hawser line broke suddenly and it collapsed limply across the pot stack on our back deck.

We could only stand by then and wait for the *City of Seattle* to sink.

For Gary Wiggins and his wife, watching from the wheelhouse of the *Cougar,* the sight of the sinking vessel was a sobering one. No one would ever know exactly what caused the ship to begin taking on water, but a lifetime of planning and working and fishing and saving had gone into building the crab boat. Nearly everything Wiggins owned was on the boat. His wife and family could attest to the endless months Gary had spent at sea. He'd been at home no more than three weeks out of the entire previous year.

"Oh, God, it's gone!" said the skipper's wife as the ship went down.

Then the bow of the *City of Seattle* suddenly reappeared on the surface. The ship was floating vertically through the ocean water, with only the top few feet of her bow showing above the surface. Her bow rose and fell in front of the *Cougar* like a whale breaching for a better look, as if to offer up some nostalgic pose or to say goodbye. Then it disappeared altogether as the ship went down for good.

Fatigued from the ordeal, and deeply disappointed at the failure of our salvage efforts, our exhausted crew fell heavily into their bunks. Without comment, our wet and disgusted skipper returned the *Elusive* to her original course down the length of Shelikof Straits.

231

FOUR

The next morning, we closed on Cape Ikolik, the southernmost tip of Kodiak Island. Several hours running time from the cape, MacDonald ordered us out of our bunks and on deck to grind up frozen blocks of herring in preparation for launching our tall back-deck load of crab pots.

Working in a rough eight-foot chop, with a ten-foot ground swell passing underneath, we off-loaded our stack of crab-pot gear. We worked hour after hour. The deck boss ran the hydraulic crane, and with the aid of the two high-strung youngsters untying the pots on top of the twenty-foot stack, each crab pot was lowered into the sloping steel rack. There, the pot became my responsibility. It was my duty as lineman to untie the crab pot's door, reach inside, and drag out the line and buoys. Then as my partner rushed to tie the door shut, I would race to separate the lines, clear any tangles, break the knot between the shots of polyester and lead-filled sinker line, and connect two more 33-fathom shots of ⅝-inch polyester king crab line. Next, I would glance to see if the two jars of herring bait had been clipped to the roof of the pot, and toss a shot of line on top of the pot itself. Finally, the deck boss would hit the hydraulic lever and the rack would tilt up sharply.

As an experienced man, he consistently managed to time this move with the starboard roll of the boat, launching the 750 pounds of steel and webbing overboard evenly and efficiently. Then, as the crab pot descended out of sight, I would toss the three remaining coils out and away from the moving boat, one by one, clearing any tangles from the line as it uncoiled on the surface, and rushing to do so before the line came tight and bound them into rigid knots or

232

hopeless tangles. Finally, I would grab the three buoys and toss each as far from our churning props as I could. Setting the gear took several hours of continuous labor. Except for the miniature rogue wave that broke over the side and filled my rain gear, and the sliding crab pot that nearly crushed one of the unwitting greenhorn youngsters, the day passed quickly and without incident.

We had hoped to put behind us the events of the previous night, but it seemed that each season had its own unique predisposition. Once the mood of the season established itself, there was little a skipper or crew could do to change it. As if to prove the point, only a few nautical miles away, the *Amber Dawn* at that very moment had begun taking on water.

Her skipper, Bert Parker, was still in his twenties when he established himself in these waters as one of the premier shrimp fishermen in all of Alaska. But due to a sudden shortage of shrimp, Parker had converted his shrimp boat into a crabber. Electrolysis, he would soon discover, had eaten the threads off the bolts holding the crab tank in place, and seawater was pouring into the ship's hull. Now, with the water level in his engine room rising fast, he wasted little time in calling for help.

"We're taking on water real bad, right now," he told the Coast Guard in Kodiak. "I could sure use a water pump! My engine room is flooding bad!"

Several hours later, we arrived on the scene and stood by the *Amber Dawn*. She was plowing ahead into the ten-foot waves like a ship made of lead. Her stern deck was only inches above the flooding waves, and as we watched, her crew raced in and out of the wheelhouse doors packing buckets of bilge water and dumping them over the side. One crewman wore nothing more than his white long johns and boots.

"My crew panicked," Parker recalls. "They wanted to get into their survival suits and abandon ship, just like that! If I'd have given the word, the entire crew would have jumped overboard." But Parker believed there was still sufficient time. The water wasn't even up to the floorboards yet, so he turned on his crew.

"You guys calm down!" he scolded them. "If we have to, we're going to bucket this thing out, at least till the pumps get here. Now go find some buckets! Now!" His sharp tone seemed to have a calming effect on the men.

A short time later, Parker managed to get one of his pumps running, and though it slowed the rate of flooding, it couldn't keep up. The water rose past the floorboard and was climbing up the stairs when a Coast Guard C-130 plane arrived on the scene. In approximately forty minutes, that plane had flown across a chunk of ocean that would have taken us sixteen hours to traverse by crab boat.

As the *Amber Dawn* rolled awkwardly through another wave trough, the black-nosed prop-driven plane roared low over her. The pilot made only one pass before zeroing in for real. On the next pass, they pushed a water pump out the back cargo door. As the package plummeted toward the sea, a parachute billowed wide, slowing the descent. Despite the twenty- and thirty-knot gusts of wind sweeping across the face of the sea, the water pump landed only yards upwind from the *Amber Dawn,* while the parachute was swept directly across her back deck. It was an impressive U.S. Coast Guard performance.

With the new high-capacity pump running, the *Amber Dawn* was soon riding high enough in the water to allow a safe journey back to the protected waters of Lazy Bay for repairs.

FIVE

We returned to our gear but were greeted by scratchy (poor) fishing, inhospitable winds, and dangerous seas. Loathing our bad luck and poor working conditions, we went to work. For the next several days, our skipper and crew went largely without sleep. Working in the rough weather, we managed to pull fewer than one hundred of our crab pots a day, while boating approximately ten thousand pounds of king crab. We had prayed for a shift in the weather, and we got it. The winds began to open up, and we were forced off the grounds and into the cover of Moser Bay. We spent the night tied to an abandoned cannery dock nestled at the base of the tall Kodiak Island mountains.

The next day, the bay was so quiet and peaceful that it was hard to believe it was blowing fifty knots outside our natural shelter. Early in the morning, a few of the boats that had been anchored nearby felt they had to go and see for themselves. They soon came back with their tails tucked, saying, "Geez! You wouldn't believe it out there!" So we waited. Unable to reach the king crab grounds and work our gear, boredom and apprehension of the strange season under way filled our days.

Each time word of a break in the weather reached us, we would race out of Moser Bay, across Alitak Bay, and begin turning over our gear. We fished in 110 fathoms (660 feet) of water, and as always, hard, sweaty work ruled the deck. But, as it was for the majority of the fleet that season, we were greeted by poor fishing and battering winds and seas that never eased. One wind-raked evening, we took time to deliver our paltry load of crab. We arrived right at sunset and anchored across the bay from the floating processor *All Alaskan*. We planned to deliver our crab at first light.

235

It was a stunning evening. Overhead hung a broad blue plate of star-studded sky, while all around us rose the sharply defined trim black silhouettes of the surrounding hills. Bare of trees, these mountains towered several thousand feet above the sea below. High overhead, their ridges flowed into one another, dwarfing the twinkling lights of the mother-ship processors anchored far below and further diminishing the forms of the proud little one-hundred-foot king crab boats tied to them.

At first light, we pulled anchor and made our way through the anchored fleet to tie up alongside Mel Wick's crab boat *Alert.* Bull-like in stature, Mel was a well-known, well-liked fisherman from Kodiak, and he and our skipper were soon together in the *Alert's* wheelhouse, jawing about the weather and the fishing.

Of the dozen or so crab boats anchored up in the bay that day, over the next few years, the *Windrunner* would catch fire and burn to a crisp; the *Gerry D*—which lost a man overboard several years before—would run aground and narrowly avoid sinking; the *Irene H* would drift onto a reef and damage herself badly; the *Kaliak* would drift on anchor and be lost to the rocks and pounding surf; and, worst of all, the *Alert,* with Mel and four other fine fishermen on board, would get caught in an ice storm and disappear without a trace.

Fishing the tough weather through, competing for the limited quota of king crab, our fleet suffered other casualties. The *Misty* caught fire and sank off Marmot Strait. It was Gary Painter and his ship *Trail Blazer* who found and rescued all five crewmen adrift in a life raft. While on her way into Alitak Bay, one crab boat, the *Periphery,* was punching her way through sixteen-foot seas and forty-five-knot winds when a plank in her wooden hull gave way. Sinking fast, the skipper hobbled her toward shore and ran her up on the beach in Lazy Bay.

The 1981 fall king crab season proved to be a long and difficult one. Some of even the most experienced crab-boat skippers found themselves halfway through the season with a catch total of less than a thousand king crab. Back in town, their grumbling crews were jumping ship at the first opportunity.

Span the six-hundred-mile gap of storm waters and rugged coast from Alitak Bay to Dutch Harbor, and there, at the very epicenter of king crabdom, one could find canneries offering only $1.35 a pound for our catch. Only? Such a price in Dutch Harbor marked a fifty percent increase over the season before, yet it lagged miles behind Kodiak's going offer now pressing close to the unheard-of price of two dollars a pound.

Only the year before, Bering Sea fishermen had harvested a record 130 million pounds of the spiny critters, and they sold them for a mere ninety-one cents a pound as fast as they could be off-loaded. Yet now, those same fishermen were scouring the same sea in search of the once-massive schools of king crab that no longer seemed to exist. Regardless of the price, hardly anyone seemed able to find the tricky red bugs.

Suddenly, it was the canneries who were doing the scrambling. They were having trouble scraping up enough king crab just to keep their people working, and many canneries were forced into week-long periods in which they sat idle and waited until crab boats blessed them with their presence.

What was going on? fishermen and cannery owners alike were asking. Was it a dark omen, a portent of the future, or some strange and inexplicable fluke of nature? Had the unending flood of king crab finally really ebbed?

When, on October 20, the 1981 Bering Sea king crab season was ordered closed and the last of the crab-boat deliveries were tallied, the total catch amounted to only 28 million pounds, a whopping 80 percent decline in fleet production in only twelve months.

Worse yet, a survey carried out by government biologists detected a dramatic plunge in the number of small prerecruit king crab (adolescent-sized crab several years growth from being legal size) out on the traditional grounds. The health of these populations had always been an accurate way of estimating the number of king crab coming of age in any one year.

With prerecruitment stocks for several years to come at an all-time low, and with the fleet catches in both the Bering Sea and Kodiak Island waters totaling only a fraction of last year's catch, a price war among the canneries began that saw the price of crab leapfrog to more than two dollars a pound!

The price rose so fast that Bill Jacobson, skipper and owner of the *Atlantico,* awoke after a night tied to the dock in Kodiak, to find that the value of the king crab he had in his hold had risen by more than fifteen thousand dollars!

Some floater processors such as the *Pacific Harvester,* the *Lady Pacific,* the *Alaskan Star,* and the *All Alaskan* anchored up on the south end of Kodiak Island, offered bait, portions of fuel, and gunnysacks full of groceries. Those floater processors who could not match the inflated price sat on anchor and waited, their cookers cold, their crews idle, and their freezers empty.

Occasionally, the demands of those with crab aboard proved ridiculous. One skipper refused to sell his king crab to a south-end processing ship unless the foreman could come up with a worthwhile amount of cocaine. When the deal collapsed of its own incredible demands, the skipper went hunting for another market.

Back in Kodiak, cannery superintendents were worried about possible consumer resistance. Longtime fisherman and cannery foreman Dave Woodruff announced that Alaska Fresh Seafoods would no longer buy crab at all.

"The price has become so prohibitive," he said, "that rather than take a loss, we're going to get out."

One Kodiak fisherman described the fishing around the island as "Generally poor. Oh, sure, it's as hot as a whore's mattress in a few tiny spots, but it's iceberg cold everywhere else!"

Yet, there were exceptions.

While his competitors scraped for a few hundred king crab per day, Bill Alwert, skipper of the powder blue seventy-six-foot crab boat *Buccaneer* was having a bumper season. He began it by delivering to Pacific Pearl Seafoods in Kodiak loads of 67,000 and 62,000 pounds of crab, respectively.

With Dirk Padrock, Larry Vandralin, and Chip Trinan working on deck, and season-end prices running well over two dollars a pound, Alwert and his crew went on to catch and deliver some 569,000 pounds of king crab. Grossing more than a million dollars, crew shares aboard the *Buccaneer* ran nearly $80,000 per man for the season, while topping $120,000 each for the year.

SIX

I had never fished with my new skipper before. Yet as the season wore on, my respect for him grew. Spot-checking for crab across the thousands of square miles of trackless waters surrounding Kodiak Island, he zigzagged back and forth along the coast, and, as with most of the fleet, did so with little luck.

Then one day off a jagged rock outcropping that shall remain nameless, he found them. The weather, too, took a suspicious turn for the better, and with less than a twenty-four-hour soak, our pots began surfacing with twenty and thirty keepers apiece. On deck, we celebrated. The kings were huge ancient-looking crustaceans. Ten- and twelve-pounders they were, stretching as wide as the hood of a Japanese import. So ungainly were they that we were forced to manipulate the rugged-looking creatures sideways through our crab chute to get them down into our live-tank hold.

For four consecutive days, it seemed as if the jinx of the season had finally been broken. For four days we turned over our gear without complaint. And for four days we earned one thousand dollars apiece per crewman each and every day. But then the dour mood of the season caught up with us again, and the Department of Fish and Game announced that the exact area in which we had struck gold would close in only forty-eight hours.

It was a moonless late-night hour when we finished hauling and rebaiting the last of our crab pots. Then unexpectedly, our skipper switched off our lights and killed our large diesel engines. Suddenly, we found ourselves alone on deck, adrift on an invisible plate of living sea, rocking in a darkness that was silent and mesmerizing.

239

"Look up" came the skipper's voice over the deck speaker. "Check it out."

Working on deck under the constant glare of the deck lights, we had not noticed. But now, overhead, was the pure white brilliance of countless stars leaping out from a coal black depth of sky, and sweeping across them danced the northern lights. It was a subtle yet delightful performance. The aurora borealis flitted and jumped in iridescent bars, like rainbows shattering in space. And from one instant to the next, they coated the twinkling heavens in transparent sheets of blowtorch greens and gaslight blues, then washed them clean with a leaping jolt of diamond white before veiling them again in rich tones of roseate red.

It was a silent and heavenly orchestration, and as we watched, the stars seemed to toss from side to side as our ship rolled gently beneath us.

It is my belief that when the mood of a season swings against a crew, the crab come hard and nobody gets along. This experience only reaffirmed that belief. All season long, squabbling broke out, often over nothing of consequence. "Who left the frying pan out?" "Who tied this pot?" "It can't be my watch! I just had my turn!" "Whose turn is it to wash dishes?" "I'm not cooking! I cooked last night. Or can't you remember?" And on and on.

Excluding our deck boss, these were the lamentations of the young and uninitiated. I'd never been around anything like it before. Their expectations— unconscious or not—of romance and overnight wealth had been exploded by the reality of a few weeks of life and labor on the high seas. That, combined with the unexpectedly poor fishing and endless work, had proven too much.

None of it came down from the wheelhouse. There was little a skipper or deck boss could do, once at sea, to make fishermen get along. He could sacrifice the trip, return to port, fire the crew, and hire another, or he could bear with it and make the best of the situation.

Finally, I came to the realization that the personalities on board ship would never blend, no matter how good the fishing or weather—not in one season, not in ten.

Pointing to another's shortcomings, however, would have done little to ease my conscience. I had no farther to look than myself to establish fault for the failed spirit of the season. For me, the glamour had gone out of fishing. I'd grown weary of the life and its endless demands, and I had acquired a certain reluctance from which boom-day Bering Sea fishermen with six-figure portfolios often suffered. With the fear of poverty and failure gone, I was no longer hungry. Part of me had become the spectator.

As we headed back into Kodiak with a payload of king crab, I felt apprehension growing within me. By the time we had secured our lines at the B & B Cannery, I had come to the conclusion that things could not go on the way they had. My stay on board the *Elusive* had come to an end.

The next day, I climbed into the wheelhouse and talked with my skipper. "Tom, do you think you can find another deckhand to replace me?"

One of the greenhorn crewmen who was dozing with a magazine in a wheelhouse chair jerked awake.

Tom MacDonald looked surprised, but without hesitation he nodded and said, "I'll work it out."

"If you can't locate someone you really want," I offered, "I'll go back out with you."

"No, I'll work it out," he repeated.

Then I noticed the microphone in the large palm of his hand. He'd been chatting with another skipper on his CB set. When he resumed that conversation, I realized that the subject was closed. My former skipper, whom I respected very much, would no doubt survive my parting. I had to smile, seeing how seriously I had taken myself. Going below into my stateroom, I packed up my belongings and moved off the ship.

Three days later, while under way some one hundred miles off the southern end of Kodiak Island, the *Elusive* caught fire and sank, very nearly taking several of her crew down with her.

SEVEN

As the 1981 fall king crab season continued, an unending series of cruel and excessive storm fronts moved across Kodiak Island and the Bering Sea, doing their best to pound the fleet into submission. One forecast called for sixty-knot winds and seas to thirty feet.

Finally, the personality of that strange and brutal season turned deadly. Call it incredibly unfortunate or blame it on what one fisherman called "the bad-luck syndrome gone berserk," but on the twenty-first of October in the Bering Sea, an incident occurred on board a crab boat that shocked the fleet. Skippered by Jens Jensen, the *Vestfjord* was coming from Dutch Harbor on its way to the gear. Jens was at the wheel and running "in twenty-foot seas and thirty-knot winds," recalls Scott Higley, a deckhand who was on board at the time that the *Vestfjord* was apparently struck by a rogue wave.

Jensen, who was well liked by many fishermen, was on the radio giving gear coordinates when the wave struck. He had his back to the wave and windows and most likely never saw it coming. "We were under way," recalls Higley, "and I felt the boat just dropping, right out of space, as if a hole had opened in front of us."

Then came the jolting crash and watery impact as the boat bottomed out. Then water began seeping through the stateroom ceilings. Higley, half-expecting the wheelhouse to be knocked back, raced up the stairwell as seawater poured down on him. When he reached the top of the stairs, Higley opened the wheelhouse door, unleashing perhaps three feet of seawater, which poured out past him, roared down the stairs, and flooded the galley. Bracing himself against the current, he continued his climb into the wheelhouse, where he came upon the body of Jens Jensen.

He was floating facedown and motionless on the floor in perhaps eighteen inches of water. A loran box torn from the ceiling by the rogue wave was floating next to him. The ship's throttle had been knocked into full reverse. No one would ever be able to determine for sure whether it was the thrust of the wave water or the skipper's quick hand that had pushed it there. Quite likely, decades of fishing instinct triggered Jens Jensen's instantaneous reaction as the boat plunged forward and the wave exploded in through the wheelhouse windows.

Crewman Higley bent and lifted Jensen's face from the water. There didn't seem to be anything wrong with his skipper; perhaps he only had been knocked unconscious. Higley lifted the skipper from the water and screamed for help. "Sig! Sig!" he screamed down the stairway. "Get up here! I need you now!"

As he held the skipper, Higley raised up and glanced around him. Then he spotted more waves approaching off the bow. With the windows gone, the wind and spray were roaring in through the gaping holes. "It was getting to a point where I was going to have to let go of Jens and try to save the boat," Higley recalls. And he yelled at his skipper. "Jens! Jens! Hey, Jens!"

When Sig and another crewman arrived, they pulled Jens into his stateroom and laid him in his bunk while Higley brought the ship around 180 degrees and began to run with the tall seas.

They tried to find a pulse; there wasn't any. Jens showed no sign of movement or life, and when they placed a small mirror in front of his mouth, there was no sign of moisture, no fog.

With their loran, radar, and radio electronics shorted out by the highly corrosive salt water, the crew of the *Vestfjord* had no way of calling for help or determining their position at sea. Some eight hours later, the *Libra* managed to locate them and guide them back to port. Sadly, Jens Jensen was pronounced dead on arrival in Dutch Harbor.

Several weeks later, on the eighth of November, out in the Bering Sea, the crew of the crab boat *Golden Pisces* was busy launching crab-pot gear over the side when the boat took a heavy roll and the stack shifted.

A young crewman named Keith Richards had placed himself between a crab pot and the handrailing in an effort to untie a pot when the deck rolled sharply. The stack shifted and a loose crab pot slid into the youngster, knocking him overboard.

The seas were moderate at the time and, turning sharply, the skipper rushed back to save the man. Perhaps only three minutes passed before the skipper was able to jockey in close enough to attempt a rescue. Yet when a life ring was thrown to within only a yard or so of Richards, the water had already numbed him to where he could no longer lift his arm to grab on.

Richards was unable to save himself. He had become "completely immobile, like a sack of potatoes, and started to sink," recalls the ship's skipper.

A crewman standing on deck panicked when he saw Richards sliding slowly beneath the surface. Clad only in his work clothes, he dived overboard and swam to his submerged crewmate. Then he, too, disappeared into the depths.

What the other crewmen couldn't see was that the rescuer had caught hold

243

of one of Richards's legs and was pulling him back to the surface. But then the would-be rescuer discovered that he, too, was being seriously affected by the numbing Bering Sea water, and he began to tire up to where he could hardly keep himself afloat. Leaving Richards then, the young rescuer began working his way back toward the *Golden Pisces.*

Once again, Richards began to sink, and no sooner had the skipper and crew pulled the severely numbed crewman back aboard than another crewman, also clad in nothing but his work clothes, dived overboard. As the second rescuer swam toward Richards, he disappeared beneath the surface. The rescuer swam to the general area, gulped air, and dived.

When he resurfaced, he gasped and yelled, "Where is he?"

"He's right behind you!!" the skipper screamed back.

The second rescuer dived again, descending more than a fathom before he managed to grab Richards and drag him back to the surface. Then he swam back to the boat with the man in tow and handed him over to the skipper and crew waiting beside the railing. As those waiting anxiously beside the railing pulled Richards up and over the side, the crewman who rescued him climbed back on board without any help at all—quite a feat in itself.

As for Keith Richards, the crew could find no vital signs, no pupils, no heartbeat, no breathing. Using heart massage and injections of adrenaline, they worked for hours to revive the young crewman. Finally, the U.S. Coast Guard doctors and medical personnel relayed a message giving permission to cease all effort. That was followed by a miserable eighteen-hour boat ride back to Dutch Harbor.

While the fleet reeled from the shock of two deaths among its ranks, only four days would pass before another tragic and far stranger incident occurred.

The crew of the *Cougar* had been fishing for king crab in the racing tidal grounds around the island of Chirikof and Tugidak, but were getting only a few hours work each day on the gear. Tides of minus 2.0 and worse were sucking their crab-pot buoys under and holding them under—out of reach beneath the ocean's surface—for days at a time.

Tired of wasting fuel, the crew on the night of November 12 said the hell with it and dropped anchor approximately a mile offshore of Sitkinak Island.

The skipper of the *Cougar* had found a fine natural anchorage in the protective lee of the hills and plateaus of Sitkinak Island. The shore was lined with cliffs four hundred to six hundred feet high, and behind that stood a series of hills some sixteen hundred feet high.

During the night as they swung at anchor, a northeast wind of twenty to twenty-five knots kicked up and a formidable due-easterly swell began moving toward the shore. The next morning, when those on board the *Cougar* stood in her wheelhouse and surveyed the situation, they decided it wasn't safe to take the skiff in and go hunting on the island. Even from their vantage point more than a mile away, it was obvious the surf was breaking too violently along the shore-line.

By noon, however, the wind had suddenly and dramatically eased and the swells had become flat. Then the sun came out, and suddenly it was a beautiful day, almost springlike, if one was to believe one's eyes and not the calendar.

A crewman on board the *Icelander* had passed word that there was some fine goose hunting to be had in the Sitkinak Island lagoon. And now two of the *Cougar*'s crew decided they'd launch a skiff and try their luck hunting ashore.

The *Cougar* had a relatively new skiff, with thirty-five-inch aluminum sides, an air compartment in the bow, and an older kicker to push it. It was considered almost unsinkable and quite seaworthy, unless something was to puncture her airtight bow cell.

Big Steve Griffing, the same monstrous (six eight and 285 pounds, give or take a meal) guy who jerked Gary Wiggins to safety during the *City of Seattle* sinking, was lying in his bunk when Kaino Dixon and Dwight Florin approached him to go ashore. "Come on, man! Let's go!" they chided him. "We're going in on the beach and go hunting."

Griffing decided to give it a rain check. "Besides," he added, "there's not enough time to give it a good hunt."

"That's fine," Kaino shot back. "Jerry's going to go."

"Kaino," Griffing argued, "the problem is not going to be getting onto the beach. It's going to be getting off of it!" He was referring to the fickle nature of the weather and the fact that it was the time of season for some of the most severe tides of the year. But there was no changing Kaino's mind. He and Dwight Florin would go ashore, while Griffing and the others remained on board.

The shore party lowered the aluminum skiff over the side and began loading it with supplies: survival suits, a 7mm rifle, a 12-gauge shotgun, hip boots, at least one flashlight, plenty of line, two oars, and crab buoys. They took no food, nor did they take along any rain gear.

Rain gear? The sun was out. It was going to be a real beauty of a day.

As they were about to depart, their skipper offered them some advice. "When you get in there, pause outside the surf and take another look," he told them. "If it looks too rough, come on back. And if you do go in, be back before nightfall. But in any way, shape, or form, if it looks shaky, don't land the skiff."

Griffing watched through field glasses from the top of the wheelhouse. He saw the two crewmen head in toward shore, pass safely in through the surf, land their skiff, drag it up on the shore, and then head up the beach toting their guns.

On board the *Cougar,* the remaining crewmen played some cribbage, cooked dinner, straightened up their rooms, and just puttered around the ship. At about 3:30 P.M., they began scanning the shore for a sign of Dixon and Florin, who should have been on their way back.

"It gets dark between four and five P.M. that time of year," Griffing recalls. Then, as nightfall neared, "a cloud bank moved in, and it began to rain."

For more than an hour, a heavy mist fell. By 4:30 P.M., they could no longer see the skiff on the beach. With the close of nightfall, the crew on board the *Cougar* began listening to local weather forecasts.

245

By 6:00 P.M., it was "low water again, and the swells had begun to come back up," recalled Griffing. Then an easterly swell rose, and the wind quickened to perhaps twenty knots.

At about this time, Griffing approached his skipper. "Mike, do you think I ought to flash a light toward the beach and see if we can get some kind of response?"

"Yah, go ahead," encouraged the skipper.

Griffing began signaling in the direction of the beach. A few minutes later, he spotted a light flashing back. It wasn't an emergency signal, just a simple tit-for-tat response. Six flashes sent . . . six flashes received.

"Mike, they're on the beach," Griffing said. "They're probably at the skiff."

Or, he asked himself, had the flashes been those of a swinging flashlight as Dixon and Florin dragged their skiff back down the beach to the surf?

A large swell was building now. The wind had kicked up and was gusting past thirty knots, marking the face of the ocean with foamy streaks.

There was nothing more to be done until daylight. The two men caught on shore by the close of darkness would be expected to remain there until morning. They would be safe there, although admittedly a mite cold and wet during the night. That, however, was part of the risk of going on such an expedition so late in the day in that kind of country. Get caught ashore by darkness or storm, and you spent the night on the beach.

The skipper fired up the *Cougar's* main engine and idled all through the night so that the ship's floodlights would illuminate the darkness in the direction of the shore area where Kaino and Florin had been sighted last.

Then at 7:30 P.M., an emergency flare shot skyward and lit up the sky. Everyone who had gathered in the wheelhouse saw the parachute flare shoot up from the beach in front of them and blow quickly off across the island. It sent shivers of apprehension through the crew.

Perhaps they had attempted to return through the heavy surf and had rolled the skiff! And what if someone was hurt—perhaps a gunshot wound. Why else would they shoot off an emergency flare? Griffing believed that something had definitely gone wrong ashore.

Suddenly, Griffing spotted what he believed to be a flashlight Mayday signal coming from the beach.

"Mike!" Griffing told his skipper. "He's blinking an SOS at us right now!"

"Are you sure?" the skipper asked, puzzled.

"I'm positive," replied Griffing. Another deckhand also had seen it.

A short time later, the skipper of the *Cougar* called the crab boat *El Dan*. Did they have a skiff on board and did it have a good running engine?

The skipper's intention, recalled Griffing, had been to get hold of another skiff and maneuver it in close enough to the shore to be able to communicate to Kaino and Florin and see whether anybody was actually hurt. If anybody was injured, they'd call the Coast Guard. If not, the men could remain on the beach, build a fire, and wait for first light.

Shortly, the *El Dan* came back over the radio. They had a skiff but no engine. Then the *Marcy J* called. They were coming out of Alitak Bay and were en route to Sitkinak Strait. They had a small Zodiac skiff with a strong running outboard motor. They'd arrive on the scene in an hour.

Those waiting in the *Cougar's* wheelhouse maintained a constant surveillance of the darkness in the direction of the shore. Sometime later, they spotted a flashlight bouncing north. It was swinging as if dangling from the end of an arm of someone who was walking.

"There's a loran station on the island," they reasoned. "Perhaps they are headed there. Or maybe they're going to hike over to the ranch house on the other side of the island."

The *Marcy J* delivered the skiff around midnight, and in the first black minutes of Friday the thirteenth, Griffing shoved off for shore. By then, the winds "were pushing thirty, thirty-five, even forty knots, and the swell was building toward ten feet," he said. What was worse, the ocean waves were pounding straight in against the steep rock shore.

For Big Steve Griffing, it was a nightmare. As he maneuvered in the wind and spray, he quickly discovered that the light rubber raft beneath him was unstable in the extreme. And so, stretching his gigantic frame across the length of its bottom, he gripped the bowline and flashlight in one hand and worked the throttle of the small outboard engine with the other.

He moved out across the water then, feeling his way in the tiny rubber dingy, lifting and falling through the steep ocean troughs with only the weak, flittering beam of a hand-held flashlight to show him the way.

But Griffing had a plan. He bucked up to the north against the wind for half an hour or so, and then angled back toward the beach, quartering with the wind and across the ever-building surf.

He did not look back for perhaps forty minutes, but when he did, he spied the *Cougar's* mast lights. The powerful beams had been filtered to a pale yellow through the fog and blowing spray. As the ship swung at anchor, her faint lights swept slowly back and forth across the bay, a body of water now disfigured by the wind and waves and by the broad black wave shadows moving past him.

The steep ocean swells were everywhere, advancing through the pale yellow light and powering off into the darkness like giant phantoms.

Griffing had intended to draw close enough to scream to the pair trapped ashore. He had planned to tell them to stay put and to remind them that there were cabins, the loran station, and the farm in the area. Then as he drew closer to the island itself, he could hear the muted crash of the storm waves spending themselves in the impenetrable darkness in front of him.

They are really stacking up along there, he warned himself. And if I get caught in that surf, it'll kill me.

The waves sweeping past him were a full twelve feet in height and measured nearly a hundred feet from wave top to wave top. They welled sharply as they neared land, and as each one passed beneath him, he rode up and over its crest, feeling all the while "the size of a particle," as he put it.

It was the exact moment in which he crested over the top of each towering

247

wave that scared him blind, for it was there that he became fully exposed to the razorlike forty-knot gusts. They struck with full force along the entire exposed length of his raft. And, each time, it nearly capsized him. He knew he had to keep moving and quartering across the waves. If one of the steep incoming rollers were to break over him, he was dead.

Suddenly, Griffing was struck with the utter loneliness of his predicament and the ever-increasing insanity of his mission. He was thinking about his wife and six-week-old daughter, Christine, when his skiff's tiny outboard motor choked and died.

With his adrenaline surging, Griffing leaned carefully over the stern of the tossing skiff, unlocked the engine, and tipped it forward. The propeller was wrapped in something. When he flashed the light on the problem, he discovered that he had drifted into a large kelp bed.

In quick succession, he cut away the binding tangle of kelp, dropped the thin outboard shaft back in place, and gave the starter cord a yank. The small engine jumped to life. He jockeyed the raft back out through the mountainous surf, maneuvering as best he could through the blowing spray and black wave troughs, refusing to give up in his efforts to reach his crewmates Dixon and Florin.

When his chance came, he turned and headed directly in toward the island, but his propeller became immediately tangled and the outboard motor died. Once again, he scrambled to cut the prop free. Now, with the engine off, Griffing found himself another three hundred feet closer to the beach. From there, he could hear the waves striking home. They were exploding headlong against the rocky shoreline, striking with the thundering reverberation of dynamite detonating. As the large swells passed underneath him, he could feel them transforming into deadly free-falling breakers.

It was then as he wrestled with the stalled engine in the pitch-darkness of the storm and surf that he heard it. There was an eerie hush as the face of the wave drew back into itself. Then, as the wave gathered momentum in its final rush toward shore, he detected what he described as a strange "hissing sound, almost a whisper" as the uppermost crest of the wave was peeled back by the stripping winds and blown into a feathery mist. And he received it as a warning, as a man who had entered into the heart of darkness.

He could feel it as much as hear it. He was near the main reef, just offshore. If only one of those steep breakers was to peak early, he knew he was history.

"The hell with this," he said flatly to himself as he tore the last of the kelp from his propeller. "There's nothing more I can do! Those guys will have to spend a cold night on the beach, that's all!" And this time when the tiny outboard cranked to life, he pointed the skiff on a sharp quartering course up and over the incoming breakers, aiming all the while for the safety of the *Cougar*.

On his way back, Griffing spotted another SOS signal blinking from the area of the beach. But he knew there was nothing more he could do. It was pure suicide to attempt to go ashore in a skiff with fourteen-inch rubber sides in seas like that. Pure suicide!

He'd go in at first light and pick them up. Whatever had happened, the men would have to see themselves through the night. It was one of the consequences of puttering around ashore on those rugged islands.

Griffing slept fitfully that night. At daybreak, equipped with food, flares, a walkie-talkie (one that could receive but not send), clothes, lights, and wearing a survival suit, he pushed off from the side of the *Cougar* and once again headed toward shore.

The weather had calmed somewhat during the night and now there remained only a light wind and a moderate surf. He'd been instructed by the skipper that if the men were there, he should light one flare. If they were not there, or if they were in need of Coast Guard assistance, Griffing was to light two flares.

He motored directly into where Dixon and Florin had beached their skiff the day before, then paused a hundred yards offshore to assess the situation. Next, he skirted along parallel to the area where they'd last been seen. Griffing made several passes in front of the mouth of the lagoon, where they had planned to hunt, but he saw nothing of the pair.

His crewmates had landed on an especially steep and rocky stretch of shore. When he approached this time, he came from the opposite end of the beach. Timing his rush through the surf, Griffing motored in and landed without mishap.

"The tide was flooding," Griffing recalled. After dragging his skiff well up on the beach, he paced off the distance from the boat to the high waterline. It was a way he'd learned to keep track of time and the danger the rising surf posed to his skiff. Then, with the raft pulled well up on the shore, he began the search.

Griffing walked hurriedly along the beach, pausing at each ravine and inlet to call out. He carried a foghorn, and as he searched, he set off loud blasts here and there and screamed aloud. "But nothing," he remembered.

Griffing had worked with fishermen all his adult life, and he strongly suspected that at any time his fellow crewmates were going to leap out at him like living ghosts and scare hell out of him. They have to be messing with my head, he thought.

It wasn't until he reached the end of the beach and stood facing the six-hundred-foot cliffs there and physically could go no farther that the impact of the situation finally struck him. He'd gone right by where the skiff should have been and found nothing. "No oars, no survival suits, no skiff, no survival-suit bags, no rifles or spent shells, no line, and no fire or any trace of them having made attempts to build one," he said.

Kaino Dixon and Dwight Florin, their skiff, and every last item of their personal belongings had disappeared. They were gone as completely as if they had vanished into the cold November air.

Perhaps they are sound asleep beneath their raft up one of these ravines, he counseled himself. It's possible the foghorn had not awakened them. He decided to take another look around.

As he searched, he passed a bundle of oblong Japanese gillnet floats washed into a pile, and a glass ball. The pounding surf and tidal currents had ripped long, flat tentacles of bull kelp from their beds and rolled them into chest-deep piles twenty-five and thirty feet long. Mounds of the slick green plants lay all along the beach, tons of the stuff. In each jumbled pile, Griffing stopped and dug through the debris for bodies.

He hiked inland but found nothing. Then he returned to the water and began zigzagging up and down the beach. Soon, Griffing came upon two sets of tracks, perhaps sixty feet of them. They were clear tracks—boot prints. They led down to the water's edge. Two people had been walking side by side. Neither print showed any sign of struggle. There was no blood, no dragging of feet, and no crutch-stick holes. From this, he inferred that neither crewman had been injured.

Again he walked the full length of the beach, blowing his air horn and calling his friends by name. Convinced that they weren't anywhere to be found, he hurried back to the raft and set off two hand-held flares.

When the skipper on board the *Cougar* spotted Griffing's message, he immediately radioed the Coast Guard for help. By 3:00 P.M., the first USCG helicopter had arrived and began searching the island. It landed first near the ranch on Sitkinak Island. There, Coast Guard searchers inquired of two ranch caretakers whether they'd seen anything of the two men lost in the area. It was the first they'd heard of the missing men, the ranchers assured them.

Over the next few days, U.S. Coast Guard helicopters and C-130 SAR planes flew seven fixed-wing-aircraft sorties and two helicopter sorties, for a total of 71.6 hours of flying time.

If the men had tried to launch their skiff and had been swept away by the currents, officially, there was a 70 to 90 percent chance of locating them, weather conditions permitting. However, if they had overturned in the heavy seas and had become separated from their skiff, the odds of locating the men floating in brightly colored survival suits hovered somewhere between 5 and 10 percent, and never rose above the latter.

In all, the Coast Guard searched approximately 37,000 square miles in and around the islands of Sitkinak, Tugidak, and Chirikof. They searched almost continuously for five days but found nothing. Finally, feeling assured that the chances of finding a living person after that amount of time were minimal, a request was made to Juneau to suspend the search, one that authorities granted.

No trace of either crewman was ever found.

EIGHT

If the deaths of men off the fishing vessels *Vestfjord, Golden Pisces,* and *Cougar* hadn't been enough to shake up the fleet during the 1981 fall season, the dramatic incidents involving the crew of the "ghost ship" scalloper *St. Patrick* were.

On the gray and wintry morning of November 29, 1981, the 158-foot fishing vessel *St. Patrick* slipped her Kodiak moorings and slid east through the narrow passage of Near Island channel and moved out into the Gulf of Alaska. For the majority of the ten men and one woman on board, it would be a journey of no return.

On board was twenty-three-year-old Wallace Thomas. A few weeks earlier, he had ventured north to Alaska from his home in balmy St. Augustine, Florida. Like thousands of naïve and adventurous hopefuls before him, he'd come in search of a berth on one of the fleet's high-paying crab boats. Unlike most, he had managed to land a job as a full-share deckhand on the *St. Patrick,* a scalloper that had arrived several months before from the East Coast by way of the Panama Canal.

That night, Thomas lay uneasily in one of the crew bunks housed forward in the bow of the ship. The *St. Patrick* had begun to plunge and leap beneath him. He could hear heavy seas crashing across the deck overhead, and each time the ship's bow buried itself in a wave, Thomas felt himself being pressed heavily into his mattress. Then as the bow rebounded, soaring high over the crest of the next wave, he would float upward, entirely free of his bedding.

Sensing that the storm was building, Thomas left his bunk, slipped into his

251

rain gear, and in the chilling spray and darkness made his way across the deck to the wheelhouse mounted astern.

It was nearly midnight when Thomas entered the comforting warmth and light of the galley. What he saw stunned him. The kitchen was in shambles. The new tightly secured microwave oven had broken free and lay smashed on the floor. The cupboard doors were swinging open and a combination of catsup, pickles, grape juice, strawberry jam, buttermilk, and sugar was sliding this way and that with the G forces of the rolling ship, in a sticky, scrambled, flowing mass. Arthur "Art" Simonton, a former logger from the state of Washington (the most experienced deckhand on board), was standing with his back to the sink, clutching its edges, as he stared at the demolished kitchen. Other crew members were leaving as Simonton turned white-faced to Thomas. "We're going to put on our survival suits. This storm is getting out of hand."

Thomas felt sick with fright. He didn't own a survival suit! And he knew he didn't stand a chance in the thirty-nine-degree seas without one. Though he had never worn one in his short career at sea, he knew about their buoyant, heat-saving qualities.

The life raft! remembered Thomas. If the ship went down, it would be his only chance! He grabbed a flashlight and ran out on the back deck, where he managed to locate the tightly packaged self-inflating raft. He read the instructions.

Looks simple enough, he thought.

Thomas made his way back inside, where he ran into the ship's newly hired cook, twenty-three-year-old Vanessa Sandin. The blond-haired, green-eyed daughter of a Kodiak salmon fisherman was carrying her survival suit, an older variety that looked in poor shape. It didn't have a built-in flotation device or life vest attached. The normally cheerful Vanessa was terrified. "Wally! What should I do?" she asked.

"The wheelhouse would be the best place to be if the ship was to get in trouble," he told her. "I'll take you up there."

Thomas climbed the stairs into the wheelhouse, then jerked to a stop. Nearly the entire crew stood before him. Most were wearing survival suits.

There was thirty-four-year-old Jack Taylor and thirty-three-year-old Curt Nelson, both from the state of Virginia. It was only Taylor's second trip out as skipper of the *St. Patrick*. Nelson was his engineer. Then there was John Blessing, a hard-working youngster from Oregon. He'd come north to help finance his college education. And there was Harold Avery, Jim Harvey, and Ben Pruitt. All three of these tough, scrappy crewmen were from Virginia. Also there at the time was Robert Kidd. This incredibly strong and sinewy deckhand was from Rhode Island. And there was Paul Ferguson, a husky lad and former football player from Nebraska. Not in the room, but also on board at the time, was a youngster named Larry Sanders, as well as Arthur "Art" Simonton.

When Thomas looked out the wheelhouse window, he saw a mountainous wave rise out of the darkness and slam heavily into the *St. Patrick*'s port side, lifting and shaking the entire ship. Wave after wave broke over the tall hand-railings and collapsed across the deck below him with a thundering crash.

252

The black foam-streaked waves looked mammoth in the far-reaching beams of the mast lights. Some of the waves towered above the wheelhouse windows, more than twenty-five feet above the deck below.

To aid vision, the light at night in the wheelhouse was always kept to a minimum. In the near darkness, Thomas turned and looked at the others. Their wide eyes were filled with fear. The faces peering out from the sealed openings of the hoods of their survival suits looked bloated.

If anything happens to the *St. Patrick,* he thought, I'll have only the life raft to save me.

Back inside, Wallace Thomas helped Vanessa Sandin slip her legs into her survival suit. Then he rushed below to look for a suit for himself. The knot in the pit of his stomach continued to tighten. As he searched below, he felt a monstrous wave strike the *St. Patrick.* The boat shifted sharply and Thomas staggered against the wall.

Thomas crossed the floor at the bottom of the stairs and started down the next gangway into the engine room. He had gone only a few steps when another wave drove into the ship. In a steady motion, the engine room rotated before him, and suddenly he found himself lying on his back on what had been the wall. Stored canned goods, oil filters, tools, and supplies fell noisily from their shelves. Several fuel lines broke and diesel fuel began to spew everywhere. Then the *St. Patrick* partially righted herself.

Thomas struggled to his feet and raced back toward the wheelhouse. His heart pounded as he crossed the sloping galley floor. The entrance to the wheelhouse stairway was marked by two full-length swinging doors. As he approached, the doors burst open and were ripped from their hinges as a wall of rushing seawater exploded through them. The broad and powerful current carried with it charts and navigational equipment. The seawater slammed Thomas against the wall. When the hallway below quickly flooded, the waist-deep water began to empty down the second stairwell, which lead into the engine room.

Even before the torrent of water had finished draining from the wheelhouse, Wallace Thomas raced up the stairs. The scene there horrified him. A giant rogue wave had smashed through the *St. Patrick's* "storm-proof" windows, tearing most of the ship's navigational equipment from its mounts. Equipment hung from the dripping ceiling, swaying from the ends of strands of wiring, while much of the ship's electronics lay broken and scattered across the flooded floor. A bone-chilling wind was gusting in through the holes where the windows had been.

The *St. Patrick* was listing about fifteen degrees to starboard at the time, making the wheelhouse floor slick and difficult for Thomas to cross. He spied his skipper lying amid the strewn equipment and broken window glass. Then he turned to Vanessa Sandin. She'd been only partially protected when the rogue wave shattered the windows and exploded into the room, and now she stood drenched and shaking.

Thomas had just finished helping Vanessa into her suit when the fuel-pressure alarm went off, clanging like an incessant fire bell. Then the main engines died and the lights aboard ship flickered, dimmed, and went dead. In the ghostly silence, a moaning wail became audible as winds approaching eighty knots howled through the steel cables of the mast rigging overhead and into the wheelhouse through the gaping window openings.

Jack Taylor found himself adrift at sea without steering or power. The only light aboard ship came from several battery-powered lamps two flights below. Then a second huge breaker drove into the side of the ship, rocking her sharply, and two crewmen were thrown to the floor. Thomas heard Taylor yell. "Hey! We've got to get off this damned thing! Let's get into the life raft before she goes down!"

An icy spray drenched them as they fled from the wheelhouse. Moving out in single file, they crawled along through the wet, cold darkness, climbing over the twisted, wave-bent handrailings and through tangles of rope and gear.

When Thomas reached the stern, he groped for the raft. "It's gone! The raft's gone!" he screamed above the howl of the wind. A few disbelieving groans met the news and then the group slipped into stunned silence. Thomas heard the skipper yell again. "We've got to get off this thing before it sinks! It'll flip over and suck us all down with her!"

A half-formed wave crashed against the far side of the St. Patrick and threw a wall of icy water over the crew. Those in survival suits paid scant attention. Clad only in his work clothes, Wallace Thomas was soaked and already ached with cold, however. He felt Vanessa grab him and heard her scream above the wind. "What are we going to do?" At that moment, Thomas spotted an amber beacon rising and falling off the stern. "It's the life raft!" he yelled. "It's got to be! Maybe we can swim for it!"

A crewman found a long length of rope and suggested that they tie themselves together to prevent getting separated once they were adrift in the ocean. The skipper agreed. Someone had located a waterproof flashlight, and periodically a voice would call out "Could you give me some light over here? I can't see!"

Suddenly, the realization of what he was about to do struck Thomas. The time twenty minutes pounded in his head. From his instruction in wilderness survival, he knew that a man without a survival suit seldom lasted longer than twenty minutes in seas of this temperature. *If you go into that water without a survival suit,* he thought to himself, *you'll be dead in twenty minutes!*

The stern deck grew steeper as the unrelenting velocity of the storm rolled the *Saint Patrick* farther on its side. Like the rest of the crew, Thomas was sure the ship was sinking. For one long moment, he stood on the stern at the water's edge, and as the water licked up the sloping deck and over his feet, he grappled with the insanity of panic. He pictured himself alone in the darkness of the wheelhouse, stretching for air as the wintery Gulf of Alaska seas rose slowly over his head.

Soon, the crew members finished tying themselves together. They were about to abandon ship, and, irrespective of logic, Thomas felt drawn to follow. As his skipper and crew edged closer to the water and prepared to jump, Thomas bolted, however.

"Taylor!" he shouted to his skipper. "I can't go in with you! I've got to stay with the ship as long as I can! I'll die if I go in that water!"

The skipper seemed dazed. "What?" he shouted back.

"I don't have a survival suit and I'm not going in that water without one! I'll stay aboard until the ship goes down and call out more Maydays!"

"I think there's another suit in the captain's cabin!" the skipper yelled back.

Thomas shook with fear as he hurried back toward the wheelhouse. He was frantic, and his legs drove him forward faster than his numbed hands could interpret the shape of things. But the suit meant life to him and he scarcely noticed the skinned shins and bruises he acquired as he stumbled along. Entering by the rear wheelhouse door, he crawled along the wall through the rubble, feeling his way to the captain's cabin.

Thomas was on the verge of hysteria. It was the prospect of being abandoned alone on board the sinking ship that terrified him. He wanted to get back to the others waiting on the stern before they became too fearful and left without him. His hands slapped frantically against the walls as he crawled through the darkness. He realized he was in the bathroom and quickly backed out. His hands shook uncontrollably as they fumbled through scattered socks, shirts, boots, pillows, a suitcase, and supplies that lay strewn about the room.

"Where is it? Where is it?" he yelled aloud.

Then as he groped in the darkness, he felt the distinctive shape of a survival-suit bag. There was nothing in it! The suit had been taken! His mind raced. Was the skipper mistaken? Could there possibly be another suit in all this junk? Had there ever been an extra suit in the first place?

He plodded ahead blindly, bumping into one wall and then another. He tore through a jumble of bedding and clothes. Reaching some cabinet doors, he jerked them open. Inside, he felt the soft bulk of another long vinyl bag. There was a survival suit inside.

Thomas wiggled into the suit and felt his way back to the stern, where he raced to tie himself between Vanessa Sandin and John Blessing. John was one of three crewmen without a survival suit. He had slipped on every piece of clothing he could wear and yet still fit into a bulky life jacket. "We've got to get off this thing!" yelled the skipper again. "She could go down any minute!"

As they prepared to jump, Thomas looked around him and caught brief glimpses of the crew in the flickering flashlight beams. Their slick, wet figures looked ghostly.

Then Chief Engineer Curt Nelson yelled a warning. "When the water hits those batteries, they're going to explode!" Just then, a loud *bang* sounded from below in the engine room. A massive wave rolled by, passing as a black hulk just off the stern. Water swept up the slanting deck and over the crew's feet.

As the stern of the *St. Patrick* dipped and swayed, the youthful six-foot four-inch frame of Art Simonton arrived. He had just returned from the bow. "Two guys just jumped overboard off the front deck!" he screamed, his eyes wide with terror. "We're going to capsize!"

"Okay," the skipper yelled. "Let's get off this thing!"

255

With the back deck constantly awash, Thomas could feel the ankle-deep water pulling at his legs. We're going down, he thought. We've got no choice but to abandon ship.

Bound together—around the waist and under the arms—by loosely tied loops for easy movement, the crew shuffled down the steep deck. The railing before them had been torn away and now, timing their move with the downward roll of the ship, the crew stepped off the stern deck.

As they struck the water, the crew went under briefly. When Thomas's head bobbed clear, he gulped in the precious air. I'm all right! he thought, his mind racing. I can breathe! I'm not dead!

In the next moment, the massive steel hull of the *St. Patrick* flashed before his eyes as it plunged down beside him, narrowly missing crushing the entire crew. Thomas screamed, "Paddle! Paddle!" and the group stroked furiously to get clear of the ship's deadly stern.

They were afloat in the stormy darkness and except for the fragile beam of a lone flashlight, vision was impossible. Looking back, only a few yards into their journey, not a hint of the ship's outline remained. Yet distinctly visible in the battering night were two small portholes. Power still generating from the engine-room batteries filled the round windows with warm light, and now the circles soared and dived in the blackness. Moments later, the porthole lights disappeared. She finally sank, thought Thomas. We got off just in time.

No sooner had the nine drifting crewmen of the *St. Patrick* swum clear of her deadly stern than a wave began to lift them. Up and up it carried them. Thomas was certain they had neared the top of the steep, sloping wave when he heard a loud thundering roar coming from behind him and far overhead. He turned toward the booming rumble just as the first monstrous wave top collapsed down upon them.

The body of the wave carried whole fathoms of sea over them. It drove the struggling crew under and tossed them end over end. Like Thomas, some of the crew members had turned toward the wave when it overtook them and caught the full force of its fury directly in their faces.

No one could have imagined such a wave. Some in the party were washed out of their rope loops. They bobbed to the surface, gasping for air, and fought to remain upright as they groped blindly for the rope.

Hundreds of feet of floating excess line had become tangled around them. And seventy-knot winds whipped an icy spray across the water, blinding those who had forgotten to turn away and stinging any face not protected with a sheltering hand.

Wallace Thomas soon discovered that two people were hanging on to him. Vanessa clung to an arm and John Blessing was hugging him around a thigh. Blessing was already shaking violently from the effects of the cold.

Thomas shouted to him.

"You're going to be all right, buddy! Just hang on!"

"Count off! Count off!" someone screamed.

Some of the crew seemed too disoriented to obey, while others were perhaps unable to hear the command.

Thomas felt another wave begin to lift him. It swept him up into the wind-torn blackness, ever higher and higher. Then he heard one end of the wave begin to roar as it folded over and collapsed through the darkness toward him.

"Look out! Look out!" he screamed. "It's another big one!"

The gigantic breaker rolled over the crew, submerging and tumbling them upside down as it passed. A few in the group had prepared for the approaching swell. Holding their breaths, they had turned their backs to it. Though they still found themselves tossed and wrenched violently about, their recovery, once the wave had passed, was surprisingly rapid. Suffering repeated dunkings, others of the drifting crew soon adapted—except, that is, for those without survival suits.

Thomas could feel John Blessing shivering violently as he clung to his leg. Less than twenty minutes had passed when his friend became delirious, moaning and speaking irrationally.

"It's cold! It's so cold!" he gasped in a painful rasping voice.

Thomas tried to encourage him. "Hang on, John! You've got to ride it out!"

But the mountainous waves continued to sweep them high. And each time the wave crests folded over, they smashed down upon the ragged group of fishermen like concrete walls.

Then one of the crewmen cried out above the noise of the storm.

"So you guys think we should pray?"

Thomas was quick to respond.

"Now is the time if there ever was one!"

They recited the Lord's Prayer then, their voices dissipating quickly in the storm-lashed night.

"Our Father, who art in heaven, hallowed be Thy name; Thy kingdom come, Thy will be done, on earth . . ."

They had completed less than half of the prayer when, without warning, another enormous wave broke directly over their heads, driving them deep beneath the surface.

The moment their heads again cleared the surface, there was heavy retching and gagging. Those tangled in the excess line fought frantically to free themselves before the next wave found them.

Wallace Thomas could see that John Blessing was in serious trouble. He wrapped his arm around the shaking man.

"Come on, John! Stay with us," he begged.

John stammered out his reply.

"My legs . . . arms . . . stiff . . . so cold."

John's moaning increased. Several times, Thomas felt his friend's quivering body go limp. When another large wave struck them, John was swept from Thomas's tiring grip and began to drift off. Stretching behind him in the wet darkness, Thomas managed to snag John with one hand. Then as he drew him near, he saw a small beacon attached to John's life jacket. He worked hard in the clumsy two-fingered gloves of his survival suit to grip the beacon and yank it alive. His aching arm muscles told him he couldn't carry his friend much longer.

The tiny beacon had just begun flashing when John spasmodically flailed his arms and floundered out of control. Almost immediately, the wind and current swept him away. Thomas was horrified to realize he could no longer reach him. He watched helplessly as the tiny amber beacon light weaved off into the darkness and disappeared.

Someone yelled, "Was that John?"

"My God, yes!" Thomas answered.

As he watched John disappear. Thomas felt Vanessa reattach herself to his arm. From the outset, Vanessa's suit had leaked steadily. It was becoming an ever-increasing struggle for her to keep her head above water. Together, they floated up and over the endless series of waves.

Vanessa Sandin prided herself in never getting seasick, even in the toughest weather. Several of her predecessors had lasted but a single trip. When Vanessa came on board, she was confronted by a galley that was dingy, dirty, and disorganized. She scrubbed the place from floor to ceiling. She emptied the cupboards and completely reorganized them. Then came the burned pots and pans, the oven interior, covered with burned grease and food, and finally the kitchen table, some twenty feet long.

She soon became known for having a good heart, and mischievous sense of humor. When Wallace Thomas had his twenty-third birthday shortly before shoving off on their fateful final journey, Vanessa had prepared him a huge birthday cake. When he tried to cut into it, however, his knife chinked to a stop. Vanessa had taken a twelve-pack of beer, wrapped it in cardboard, and camouflaged it with a thick spread of canned frosting.

She could also cook. Each summer, she fished right along with her father on his gillnet boat in Bristol Bay. One day, when she had decided to prepare a dinner of sockeye salmon, her father, a long-time Alaskan fisherman, made the mistake of trying to tell her how to go about cooking it.

"You run the boat," she scolded him, "and I'll fry the fish. Now get out of here!"

The salmon proved to be the best he had ever eaten.

Vanessa, her father recalled, could not only cook but also tie knots, mend webbing, make a drift set, pick tangled salmon out of the gillnet, and navigate. One day, it came to him that she learned more quickly and with greater ease than he had as a young man. But now this adventurous young woman could only cling to Thomas through the night and pray to be rescued at first light.

As they passed over the top of another immense swell, Thomas felt something bump heavily into his back. Fearful that a log had drifted up on them out of the night, he shouted, "Get that light over here! Something just hit me!"

Thomas searched the darkness with the dim beam of light. Suddenly, he reeled in the water. There, adrift beside him, floated the body of Larry Saunders, another crewman who had abandoned ship without a survival suit. He was still tied to the crew's rope line. "Turn off the light and save it!" shouted Thomas. The light flicked off.

Close by, Thomas overheard the muffled conversation between Ben Pruitt and Jim Harvey. Harvey had been on his first voyage to sea in Alaska. He was the last of the original three without survival suits. Now he turned numbly to his crewmate and pleaded.

"Ben! Ben, I'm so cold! Could I borrow your suit for a little while?"

"Jim, I'm cold, too! And the suit's leaking, anyway! Just try to hang on!"

"Ben! What am I going to do? Help me! Please, help me, Ben!"

Thomas knew that there was nothing anyone could do. Hypothermia was the dangerous lowering of one's body temperature; more specifically, the temperature of one's core, comprised of the heart, lungs, and brain. The moment the three frantic young men jumped from the stern of the *St. Patrick* without survival suits, they were doomed. Under such conditions, no man could live more than an hour.

A short time later, Jim Harvey began to moan and jabber incoherently. Eventually, he grew motionless and, still bound to the group by the rope line, his body drifted amongst them.

Eight hours after they had first abandoned ship, dawn slowly replaced the smothering black veil of darkness. In the dim gray light, Wallace Thomas could finally take in the unbelievable size of the massive seas. Raw winds whipped thin white streaks of foam across the moss green water. The waves moved under a bleak ceiling of sky, and he could make out blue-black rain squalls, squatting low as they moved across the horizon.

Each time he passed over the crest of another wave, his eyes swept the desolate expanse of water. Then something caught his eye.

"I can see somebody swimming over there!" he yelled. The six remaining crew members soon spotted him, too.

"Hey, it's the skipper!"

They yelled and waved and blew metal whistles that came attached to their suits, but Jack Taylor showed no sign of having heard them. He backstroked slowly away from them and was lost from sight.

Then Vanessa began to yell excitedly.

"I can see land! I can see it! I'm sure of it!"

As he crested another twenty-five-foot wave, Thomas caught sight of it, too. Before long, he was able to make out two separate points of land. The spirits of the numb, pain-racked crew soared.

"When I get back," promised Vanessa, "I'm going to eat the biggest pizza I can order!"

Hours later, when the low-lying clouds cleared adequately, a steep rock coastline loomed large before them. Its uppermost slopes were covered with rich shades of winter-browned grass and crowned with dark green stands of spruce trees.

The crew members decided to paddle for the nearest outcropping of land, but wave after wave struck them from the side, throwing them off course, and fog intermittently obliterated all sight of land.

"We're never going to make it this way!" Thomas called out. "We've got to swim more in line with the way the waves are moving! Then we can angle a little bit at a time in toward land!"

259

Swimming with the waves, they made steady progress, but soon Ben began to speak in fragmented phrases and suddenly collapsed facedown. The two crewmen on either side rolled him over.

"Come on, Ben! You've got to swim!" they screamed at him. "You've got to help us! We're not going to make it if you don't swim!"

Ben Pruitt tried. He flopped one weak arm and then the other out in front of him, but he was nearly unconscious. His two friends continued to encourage him as they pulled him along between them, kicking their numb legs and stroking with one hand while clutching Jay by the arm with the other.

Vanessa, too, was nearing the point of exhaustion. With her leaking survival suit now nearly full of the icy seawater, her body ached with cold. If she was to survive, she would need constant help.

Then Wallace Thomas and Harold Avery devised a method. One of them would paddle on his back, carrying Vanessa on top of him, while the other trailed the pair, watching for land and verbally guiding them.

Paul Ferguson and Curt Nelson soon adopted this technique to carry Ben. Gradually, they grew weaker and their progress slowed appreciably. Their bodies were chilled from nearly fourteen hours in the near-freezing water. Hunger cramped their stomachs and they were becoming dehydrated.

The survivors had decided to save the bodies for decent burial. But then, under the sinking weight of the body of Larry Sanders, Thomas saw how Ferguson was struggling to remain afloat.

"We've got to untie him!" he yelled.

Thomas swam forward through a tangle of floating lines and loosened the rope line. The corpse quickly sank from sight. Without comment, the six remaining crew members regrouped and pushed on.

Ever since they had first spotted land that morning, the struggling crew had been swimming. Now, some eight hours later, the coastline seemed much closer. Thomas could see the black flat faces of cliffs, perhaps a hundred feet high, lining the shore, but he could not see an accessible or calm stretch of beach. As they closed to within perhaps two miles, Thomas spotted the white explosions of waves bursting along the cliff bottoms.

The sight petrified him. He knew the deadly power of coastal breakers from years of surfing in Florida. Such seas were frightening enough, but to become entangled with one another in the rope lines in a heavy breaking surf would be suicidal.

"Before we try to go in, we've got to untie!" he called out to the others. "It's too dangerous! We won't make it this way!" The exhausted, floundering members numbly agreed.

Now Ben Pruitt seemed on the edge of collapse. The two who had carried him no longer appeared to be able to do so. Thomas left Vanessa with Harold Avery and swam to Jay and held his blue face out of the frigid water. He was still breathing.

Then off to his right, Thomas caught a flash of something white on the water. As it rose over the crest of the next wave, he saw it clearly. "There's a ship!"

Those who were able began to wave wildly and blow their whistles. They swung their shivering arms back and forth till they could no longer hold them aloft. Thomas wore his tongue raw whistling and the taste of blood filled his mouth. The ship closed toward them for nearly a half hour. Wallace and the rest of the crew members were sure it was coming for them.

Then they watched in disbelief as their rescue ship began to turn away. As it changed course, Thomas could see the distinct lines of the ship's wooden hull.

"No! No!" he cried. "Why can't they see us? Please, please see us!"

But the vessel was soon lost from sight.

The collective disappointment was almost too much. Wallace Thomas was the first to break the silence.

"We've got to get swimming again, you guys. He didn't see us."

With disappointment showing in every stroke, the remnant crew once again began to plod ahead toward shore. They had hardly begun, however, when Ben Pruitt rolled facedown in the water. Paul Ferguson and Curt Nelson summoned all their remaining strength to roll him back over and give him mouth-to-mouth resuscitation, but there was no response.

"Come on, you two!" Thomas finally called, his voice breaking. "You've got to let him go now. You've got to save yourselves. We've done everything we could. We've got to take care of the living now!"

The remaining survivors of the *St. Patrick* were too exhausted to untie Ben's body, so they left it in tow and resumed swimming.

Less than an hour later, the five remaining survivors closed to within what they believed was a half mile of the shoreline. "I'll go on in," yelled Harold Avery, "and if it's all right, I'll wave for you to follow on in after me! I've got some waterproof matches and I'll get a warm fire going. So just watch for my signal and then follow me in!"

Thoughts of a crackling-hot fire lifted the spirits of the four remaining crewmen. Vanessa was shaking constantly from the cold water seeping into her suit. She was growing visibly weaker. With Vanessa lying across his lap, Thomas shuddered as he watched Harold swim away.

They were able to stop paddling then and, sighing with relief, settled back to await the signal from shore. As they drifted nearer to shore, they decided to untie themselves from each other in preparation for the swim in through the surf. Freed from the rope line, Vanessa and Thomas found themselves drifting away from Paul Ferguson and Curt Nelson. But the two couples were too fatigued to reunite.

Shortly, Thomas saw the two men drop into a hollow in the sea and disappear. That afternoon, he caught his last brief glimpse of them. They appeared to have stopped swimming.

Now, with Vanessa completely dependent upon him for her survival, Thomas drifted and waited in anticipation for a signal from Harold Avery. Each time, as he rose up and over another wave, he would search for his good friend and deckmate's wave from the distant banks. Drifting ever nearer, he studied the soaring rock precipices. He could see the ghastly black form of the cliffs, slickened with spray and rising abruptly from out of a pounding misty-gray surf, with fog rolling across its sheer granite face.

He contemplated that perhaps he had underestimated the size of the surf and cliffs along the shore.

Suddenly, he caught the flicker of something tiny and orange in the thundering surf. It looked toylike, about the size of a petite orange buoy as it was lifted up and tossed against the wet face of the rock cliffs. He watched it being swept out by the surf, only to be gathered by another massive wave and flung high against the stone walls.

Then as he drifted nearer, the true dimensions of the terrain ahead finally struck Thomas. The rock cliffs were not a mere one hundred feet high, as he had estimated, but in frightening reality towered more than a thousand feet overhead!

When he spotted the minute orange object again, it was suspended in a wave and being swept some thirty feet up the face of the cliffs, with spray exploding far above it. And at that moment, it dawned on him—the object he had been studying so intently as it surged back and forth against the cliffs was not a buoy. It was the lifeless body of Harold Avery.

The sight took his breath. His heart felt like lead. All seemed lost and utterly hopeless. The safety they had associated with the first sighting of land had been only another cruel illusion. Vanessa hadn't seen the body. Thomas decided not to tell her.

"We can't get in here!" he yelled to her.

But what if there isn't any accessible beach on this island's entire shoreline? he worried secretly.

As Vanessa lay across his lap, Thomas noticed that her condition was worsening. She could no longer move her legs, the feeling had gone out of her arms, and her lips had turned a dark blue. As he paddled on his back over the waves, he tried to parallel the coastline.

He, too, had begun to shake uncontrollably, and his legs felt stiff and weighted. Toward evening, Wallace Thomas's back and arms began to cramp badly. He felt he couldn't carry Vanessa much longer. As he lay back, he thought he'd close his eyes and doze for a moment. It seemed only seconds before Vanessa rattled him awake.

"Wally! Wally, are you all right? You're looking pretty bad!"

"I'm fine, Vanessa," he reassured her. "I was just trying to relax for a few minutes."

Then a loud thumping noise began to pound in their ears. As if out of nowhere, an orange and white helicopter roared past them, close by overhead. In an adrenaline-pumping rush of excitement, Thomas waved his arms wildly; Vanessa raised one quivering hand.

"It's the Coast Guard! My God, they've finally come for us! They knew we're here! We've done it! We're going to be rescued! We're going to live!"

The helicopter sped out of sight, though. A few minutes later, they saw it making another pass in the distance.

Vanessa was too exhausted to wave.

"Did he spot us?" she asked in a weak voice.

"No. And he's too far away now."

He'd hardly finished speaking when another helicopter flew directly over them. As it passed, the side door slid open and Thomas could see a crewman standing in the doorway. He appeared to be looking right at them.

Thomas tried to rock forward, thereby rising slightly in the water, but Vanessa lay heavily across his lap. He screamed and waved frantically.

"Come on, Vanessa! We've got to signal to them! This might be the one time they see us!"

"I don't think I can anymore, Wally," she replied weakly.

The helicopter flew on out of sight.

"They didn't see us, did they?"

Thomas answered with silence.

"I don't think I'm going to make it, Wally," said Vanessa, her voice straining with pain and fatigue.

"Come on, Vanessa. We can still get out of this mess alive. We can float a good while longer if we have to."

"Oh, Wally," she replied in a disheartened voice, "I don't know. I'm awfully cold and there's a lot of water getting into my suit."

"Look over there," argued Thomas. "See that point of rock? Look past it. The waves aren't even breaking. There's a cove. We'll swim in around there somewhere."

Suddenly, Vanessa began to cough roughly. Then she jerked forward out of his arms and rolled facedown in the water.

Thomas was horrified. He jerked her back upright and shook her violently.

"Vanessa! Wake up! Say something to me! Answer me!"

There was no response. Her face was chalk blue. Her eyes were glassy. She hacked deeply then and again wrenched free of Thomas's grip. Summoning what little strength he could, Thomas paddled to her side. He lifted her head and held her close. Her eyes were closed. Her body hung limply in his arms. She was no longer breathing. Vanessa was gone.

Thomas turned and slowly swam away. He was weary and heartbroken, and darkness was closing fast. His entire body ached with cold and now shook uncontrollably. He had been awake for more than thirty-six hours, twenty of them spent battling the stormy Alaskan seas. Now he wanted only to close his eyes and be done with it. His movements had grown sluggish to the point of immobility, but he fought to keep his tired mind on the task at hand.

I'm dead! I'm going to die! he concluded.

Shortly after nightfall, Wallace Thomas spotted a ship's mast lights. He did not grow overly excited. The lights were miles off. Each time, as he rose over another wave top, he caught glimpses of them. They appeared to be headed in his direction.

As the ship drew closer, Thomas tried to gather his failing strength. He lifted his leaden arms and began waving, drawing an arm back down to rest now and again. Occasionally, he called out, hoping his voice would somehow carry to those on board the ship. Then the vessel pulled up even with Thomas, and he screamed, "My God, I'm here! I'm right here!"

He fumbled for his whistle and began blowing it frantically. Then he held his breath and listened for a reply. He could hear the sounds of men's voices above the low rumble of the diesel engines. He could see the figures of crewmen working outside under the back deck lights. Yet the ship slowly lunged past him and disappeared into the night.

I'm dead, a goner. I'm going to die, he thought.

He struggled to come to terms with the finality of his predicament. He thought of his parents—how sad and wasteful losing their son this way would seem to them.

I'm sorry, Mom! I'm sorry, Dad! he thought, picturing them now in his mind. Such a lousy way to die, he pondered.

Thomas's shuddering body throbbed with cold. Then as he rode over the crest of a wave, he spotted tiny lights flickering in the distance. Only a few hours before, Thomas would have known that they represented another ship miles away, but his thinking had become disoriented. He was sure they were the lights of Kodiak. "I'll swim in there," he decided. But seconds later, he'd forgotten the idea and paddled numbly ahead.

Thomas knew he had to get out of the water. His body was just too cold to remain in it much longer. Several times, his legs grew so stiff that he was sure he no longer could use them. He knew he would soon pass out from the relentless cold. His wilderness training in hypothermia flashed through his mind. If you got cold, you didn't try to ignore it, he remembered. You act!

He crouched up into the fetal position then, and as he drifted in the darkness, he tucked his mouth down into his suit and for a time breathed heavily into it. His debilitating numbness and shivering seemed to diminish slightly.

Then, as he came off the peak of a big loping wave, Thomas saw something large in the moon-tinted darkness. It was floating beside him. It looked like a huge buoy, partially covered with dark green blotches of algae or seaweed, but its dimensions puzzled Thomas.

Maybe I can hang on to that and get some rest, he reasoned. I'll just float along for a little while.

He paddled toward the object, but oddly, he didn't seem to be getting any closer. A moment later, Thomas drew back, frightened of the thing. Now it looked like a whale, and, retreating, he splashed water at the massive creature in an attempt to frighten it away.

Eventually, Thomas came to realize that the whale he had feared was actually a point of land less than a half mile away. Maybe it's not your time, he thought hopefully. You've got to at least try to swim for it! At least you can do that! He turned then and using the breaststroke headed in toward land.

"It's not your turn. It's not your time," he chanted to himself. "Going to die if you stay out here any longer. May be the last thing you ever do. Might as well swim for it. Got to try."

Shortly, Thomas found his movement impeded. A swaying tangle of slimy fingers was bobbing about him, while others wrapped themselves around his arms and legs. Fear began to build. Oh, kelp, he realized suddenly. Must be getting closer. Must be.

He struggled on toward the shore, and on either side of him, in the faint light, he spied tall pillars of blue-gray rock and moon-silvered breakers fanning out and bursting high against them. When he heard the roar of the breakers exploding along the shore, he grew sick with fear. The vision of Harold Avery's body as it washed up against the cliff kept shooting through his groggy mind. Aim yourself in between those two pillars, he told himself. It's your only chance.

Wallace Thomas had no sooner decided upon his new course than he felt an immense wave pick him up and hurl him forward through the night. The wind blew sharply in his face and the water churned beneath him as it heaved him along toward shore. Then, without warning, the wave crest he was riding curled forward.

He felt suspended in air as he fell down its folding face. When he landed, the tremendous force of the pounding water shoved him under the surface and held him there. The boiling torrent of ocean surf pulled and pounded on him as if nothing short of his total destruction would satisfy it. It twisted him upside down, jerked him sideways, and rolled him about. The smothering black surf seemed to pull at his suit from all directions, and he could feel icy rivulets of water jetting in around the facial opening of his hood.

Thomas fought to right himself and return to the surface. His lungs burned for air. He had already begun the involuntary inhalation of the icy salt water when his head finally cleared the surface. Thomas choked violently and gasped in a lungful of the damp sea air. Then another huge wave caught him and once again launched him swiftly forward through the night.

He threw his battered arms out in front of him and attempted to swim along with the thrusting power of the wave. Then he thought he felt a hand strike a rock, and the foaming rush of water that had carried him there seemed to disappear from beneath him.

Thomas found himself lying facedown on the steep face of a solid rock bank. He clung to the steely cold surface in disbelief, his chest heaving for air. He felt too weak to move, but he knew if the next mammoth wave was to catch him still lying there, it might very well crush him with a single paralyzing blow.

In his mind, he stood to run, but his legs refused to move. It was as if they were no longer part of him! "Oh God! Dear God, help me!" he cried, pawing wildly at the slick bare rock of the bank.

Thomas had managed to crawl only a single body length up the surf-slickened bank when the next wave exploded at his feet, drenching him.

A bitter cold wind was gusting along the shoreline and soon it chilled Thomas to his core. The short stretch of rock he had lucked upon was only a few yards wide. Too weak to stand, and shivering uncontrollably, he pulled himself along with his hands, and, lucking upon a shallow rock crevice, he instinctively rolled into it. There, out of the direct assault of the wind, Thomas closed his weighted eyes and almost instantly fell asleep.

Yet it was a fitful rest. The surf pounded loudly only a few yards away, and even in the naturally protected chasm, the razor-sharp wind found him.

Wallace Thomas awoke with a start, to see the figure of Art Simonton standing close by and staring at him. His former crewmate and friend wore street clothes and seemed unaffected by the arctic wind racing along the shore. Was it the visible apparition of a dead and departed friend, or had he survived?

"Art!" he called out. "What are you doing here?"

When no answer came, Thomas took a moment to reposition himself and draw closer. But when he looked again, the figure had disappeared. He slumped back down and slipped into unconsciousness, lost in a merciful slumber.

The excruciating pain in his hands awakened him next. The arms of his suit were bloated with salt water forced in by the pummeling surf. He rolled onto his back, lifted his arms, drained the stinging water into the lower half of his suit, and fell back to sleep. When he awoke later, he found his hands had warmed to a point where he could at least open them.

The Coast Guard was finally able to verify that twelve crewmen had been aboard the fishing vessel *St. Patrick*. At first light, U.S. Coast Guard helicopter pilot Lt. Jimmy Ng lifted off from his base on Kodiak Island. Along with several other helicopters, C-130 SAR planes, and the Coast Guard cutter *Boutwell*, they began searching the area off Afognak Island for signs of survivors.

Lt. Ng worked his way around the steep rock shoreline of Marmot Island (positioned approximately four miles from the shores of Afognak Island). Some of the shoreline cliffs he encountered rose up in a sheer vertical climb more than twelve hundred feet above the water. A short time later, Lt. Ng located the first body. It was floating facedown in a small cliff-encircled cove. Oddly, the man wore neither rain gear nor a survival suit. The cliffs were too high and the cove too small to maneuver safely, so Lt. Ng hovered approximately fifty feet away and watched for signs of life. There were none, so he resumed his search around Marmot Island, and he soon came upon several more bodies.

"One body was off in the surf," he recalled. "He was bouncing around in the rocks. And two others were lying up on the beach. All were dead."

With the discovering of the first body, the Coast Guard search for the missing crew of the *St. Patrick* intensified. Soon, USCG helicopters, planes, and cutters were scanning the waters and shoreline from Whale Island, just offshore from Kodiak, to Marmot Bay, to Iszuit Bay, and completely around the northern tip of Afognak island, all in the hope of finding someone still alive.

Shortly after dawn, Wallace Thomas propped himself upright on a boulder. He sat shivering and studied the world around him. Overhead, rock walls rose as sheer and apparently inaccessible as those of a prison. More than one thousand feet above him, Thomas could see convoluted outcroppings of bare granite rock jutting into the sky, and beyond that, clinging to thin layers of soil, weather-stunted spruce trees bent in the wind. On either side of him, short stretches of narrow shoreline cut into the cliff rock and were strewn with boulders the size of dump trucks.

The cloud ceiling appeared to have lifted slightly but the freezing thirty-knot winds continued to blow without pause. The sea rushing up at him wore a blinding silver sheen. His pain-wracked legs still refused to support him, so he rubbed them furiously in an effort to restore circulation.

Thomas felt groggy and exhausted, miserably cold and hungry. But it was a maddening thirst that drove him finally to rise on wobbly legs and stagger stiff-legged along the cliff bottoms in search of fresh water. With water sloshing about inside, his survival suit hung heavily on him. He wanted to shed the suit but knew that to stand exposed to the Siberian-born winds in soaked clothing would mean death within hours.

Maybe I'll make it if I stay in the suit, he reasoned.

Thomas could find no fresh water close at hand. Walking only a few steps exhausted him.

If I'm going to survive, I've got to locate water, he told himself.

But there appeared to be no escaping the cliffs lining the beach. The sharp-crested outcroppings of rock extended well out into the surf. They loomed impossibly steep and dangerous to climb.

The pounding surf before him, which had taken the lives of so many of his companions and nearly his own, now petrified him. He would remain trapped on the shore and take his chances with hunger and thirst and exposure before he would return to the ocean again.

Then as the surf receded to near-dead low tide, he saw an opportunity. The tide had receded enough to allow passage around the base of the jutting column on his left. Wallace did not hesitate. Leaning against the rock walls, he hurried around them as quickly as his buckling legs would carry him.

He discovered an even shorter stretch of enclosed shoreline. Large boulders covered most of it. He stumbled forward and fell on the bank, panting. Slowly, a faint dribbling sound struck his consciousness. He spun and his eyes caught the movement of a tiny stream of water trickling off the face of a vertical rock bluff.

He rose staggering and fell. He crawled the last few feet but found a small pond formed where the droplets had landed. He dipped his glove-covered hands anxiously into the clear pool and sipped the bounty. The water tasted salty and he spit it back out. His heart fell.

Damn! A tide pool! he thought angrily.

Then he sampled the pool again.

It *was* fresh water!

Thomas felt foolish. The salt he had tasted had come from the gloves of his survival suit. The water in the pool was fresh and cold. He drank down a few eager swallows and then stopped abruptly. He could feel the water cool him and he wanted to allow his body time to catch up.

Even with fresh water, Thomas doubted he could make it through another night in his wet clothes without food or a fire. It had been nearly forty-five hours since his last meal and he was sick with hunger.

In an effort to hide from the painful and life-sapping cold of the December winds, Thomas hunched down between two huge boulders. From there, he could still command a view of a good portion of the ocean. As he waited, he shook so hard it felt as if all the bones in his body were rattling. Gradually, he thought he could make out the faint rumble of an engine. The noise dimmed, then grew stronger, only to fade once more. He debated whether to stand and look, exposing himself fully to the draining cold of the wind.

He had to try, he decided. His body was fast growing colder. He wavered as he stood, and his eyes squinted into the wind and swept quickly over the ocean before him. He was about to crouch back down when he spotted movement. It was the bow of a ship nosing its way through heavy seas off a point of land on his far right. Almost crippled with cold yet frantic with excitement, Thomas struggled up the side of a large boulder and began flailing his weary arms.

"They must be out looking for me! They've got to spot me! They've just got to see me now!" he cried out loud.

The one-hundred-foot ship drove nearly halfway across the open stretch of water in front of him before Thomas thought he saw it slow. Through watering eyes, he saw quick flashes of light coming from the wheelhouse.

Were those really signals? Have they actually spotted me? Or am I only imagining things again?

Every few seconds, the ship would crest the top of another swell and then slide into a deep trough, and then, except for the radar scanner spinning steadily atop the wheelhouse, it would disappear as if swallowed whole.

The waving seemed to take the last of Thomas's strength. A sudden gust of wind staggered him, nearly toppling him from the rock. He dropped to his knees to maintain his balance. If they did see me, what will they do next? What could they do? his fuzzy mind puzzled.

On the horizon, Thomas caught sight of a small black dot moving directly toward him low over the water. Moments later, a four-engined C-130 U.S. Coast Guard plane roared by overhead. Its deep, growling engines shook his insides like the blast from a cannon. Thomas was ecstatic. He blew kisses and screamed excitedly. "Yes! Yes, they've seen me! Thank God, at least they know where I am!"

Soon, he spied a U.S. Coast Guard helicopter moving toward him. "I'm here! I'm alive!" he called out.

The wind was blowing hard against the one-thousand-foot cliffs behind him. The helicopter flew in twice over Thomas and hovered, only to clack noisily away.

Dear God, he can't get to me! he thought. I'm too cold! I've got to get out of here!

Then the bright orange, black, and white helicopter returned and hovered not fifty feet above him. The 110-knot wind churned up by the copter's blades whipped a mist off the water.

Thomas was thrilled when he saw the large steel body basket descend from out of the side door. But it landed well out in the breaking surf, and the pilot seemed reluctant to move in closer to the cliffs.

Gradually, the helicopter pilot maneuvered the craft closer, resting the basket in the surf on the edge of the shoreline. Though terrified of the water, Thomas shuffled down the embankment and fell into the basket. Fear that the next wave would catch him and batter him to death now that he was so near to being rescued raged in his mind. Yet almost instantly, the helicopter plucked Thomas up and out of the surf.

Thomas watched the shoreline grow minute in the distance. A cutting wind whipped over him as the upward acceleration of the helicopter pressed his body hard against the wire-meshed basket's bottom. Then the helicopter leveled off and stood away from the shore, hovering noisily. Thomas could feel himself being lifted toward the door.

Far below, he could see the ship that had first spotted him and radioed his location. It was the *Nelle Belle*. She was throwing off heavy sheets of bow spray as she plowed through the waves.

The helicopter crew hoisted him in through the door and checked him for serious injuries. Next a radio headset was fitted over his head, and Thomas heard the voice of the chief pilot.

"Are you all right, young man?"

"Yes. I think so."

"How do you feel?"

"Well, you got me out of there! I feel so much better now."

"Look, you've got hypothermia! Do you understand that?"

"I kind of figured as much."

"Okay, so now listen to what I'm telling you! Do not relax! Keep yourself charged up until we can get you into the hospital. You could go into shock right now and you could die before we could get you there! That has happened to us before."

The helicopter pilot was worried about a dangerous phenomenon called "after drop," the process in which a hypothermia victim's core temperature continues to plunge even after he has been rescued and wrapped in wool blankets. If not halted, this downward slide will continue until the victim suffers a heart attack from the cool blood circulating through his heart.

"Have you found anyone else besides me?" asked Thomas.

"No. You're the only one so far," answered the pilot.

The somber news shook Thomas. He had hoped that Paul Ferguson or Curt Nelson had somehow found a way safely ashore. He wondered about Bob Kidd and Arthur "Art" Simonton, whom he'd seen in his dreams the night before. Bob Kidd had jumped overboard with Simonton from up near the bow minutes before Thomas and the rest of the crew had abandoned the *St. Patrick* off the stern.

When Wallace Thomas arrived at the hospital, his body temperature was ninety-three degrees Fahrenheit. Death can occur from heart failure at ninety degrees. The medical staff placed heated blankets and hot towels across his body and forced him to breathe heated oxygen. But it took little coaxing. The warm devices felt wonderful to Thomas's numb, sea-ravaged body.

The next day, Thomas learned that one other crewmate had survived the ordeal. As he lay recovering in a Kodiak hospital bed, nurses wheeled in Bob Kidd for a visit. "I can't believe it," Thomas finally confided to his good friend.

269

"I would never have believed that a ship built like the *St. Patrick* could have gone down as quickly as she did!"

Bob Kidd sat upright and turned and looked at Wally Thomas in astonishment. "Wally," said Kidd, "it didn't go down. It didn't sink. They found the *St. Patrick* floating the day after we abandoned ship. They're towing it in right now!"

NINE

The stories of death and deprivation during the fall season of 1981 went on until one became stupefied by the unrelenting slaughter. It seemed even more impossible that in a high-tech civilization that boasted fingertip convenience, built-in safety factors, and government regulations and safety codes in nearly every trade, an occupation so dangerous as Alaskan crab fishing could have gone virtually unnoticed.

Yet during this time, crab boats working the waters of the Gulf of Alaska and the Bering Sea were not required by law to carry either boat insurance, survival suits, emergency locator beacons, or life rafts, though many crab-boat skippers did. One marine surveyor told me that in 1981 alone, 15 percent of all crab boats fishing in Alaska were either "destroyed, sunk, salvaged, or, in some way, had to be repaired." Understandably, Lloyd's of London's insurance premiums nearly tripled.

Over the next several years, the recital of death would flash across the newspaper headlines until one's heart became numbed to the pain and eventually hardened against it: BOAT SINKS—TEN DIE, SHIP REPORTED OVERDUE, SHIP WITH CREW OF FOUR MISSING, SEARCH CONTINUES FOR MISSING CRAB BOAT.

Too many of us were dying, and no eulogy or insurance settlement could compensate the friends and families left behind for all their pain and sorrow. Talk with parents of those lost without a trace. Meet their eyes, if you can bear it. Listen to what they say.

"I always find myself saying Why? *Why? WHY?*" one parent told me. "We're not naïve people. We've been in this business for three generations and

271

are aware of the dangers. But there is no way to get used to it. You deal with it every day."

When blended with the literally thousands of boats involved each year in all other commercial fisheries in Alaska (there are twenty-six different permit areas for salmon fishing alone), the fatality rate for fishermen in Alaska is about twenty times as high as the average of all other industries within the state. But when one focuses solely upon the comparatively few king and tanner crab ships that composed the fleet during these years, the rates soar to levels unequaled in any labor anywhere.

From September of 1982 to September of 1983, approximately sixty-eight commercial vessels sank in Alaskan waters, claiming the lives of some forty-six fishermen. Thirty-six of these crewmen were killed aboard crab boats.

Nineteen eighty-three, for instance, was as grim a year as anyone could remember. February started the slaughter, when two crab boats, the *Americus I* and the *Altair II*, capsized and disappeared only an hour or so out of Dutch Harbor. Fourteen fine young men, most from the same small fishing village of Anacortes, Washington, died in a single day.

Over the next few weeks, the crab boats *Flyboy, Arctic Dreamer*, and *Sea Hawk* went down, taking another three lives with them. A month later, the *Lou Anne* rolled over and sank, taking five crewmen down with her. In August, not far from where the *Americus I* and *Altair II* went down, the crab boat *Ocean Grace* foundered and sank, killing another four crewmen. In September, the *Endeavor* rolled over and all four crewmen aboard vanished with it. Thirty lives were lost in a matter of weeks.

In the next few months of 1983, even more bad news broke for the Alaskan king crab fleet. It came in the sudden and undeniable reports that the king crab stocks around Kodiak Island and in the Bering Sea had both completely collapsed.

In two years, the catch totals for the crab-boat fleet working the waters of the Bering Sea had plummeted from a high of 150 million pounds in 1980, to 30 million pounds in 1981, to a 1982 catch of just over 10 million pounds. Dutch Harbor dropped from the number-one port in the nation to number twenty. After weighing the evidence of extensive summer surveys around Kodiak and the Bering Sea, the Alaska Department of Fish and Game canceled the 1983 fall king crab season altogether.

In surveys taken out on the traditional king crab grounds around Kodiak Island, approximately 60 percent of the females trapped proved to be barren. Equally troubling was an almost complete absence of young prerecruit king crab, a sign that practically guaranteed that future crab populations and catch quotas would be unusually low, virtually eliminating the traditional Kodiak and Bering Sea king crab seasons.

Many old-timers who had fished in those waters dating back to the pioneer days knew that the king crab boom couldn't last forever. Nothing stayed the same for long in this business, especially in such a cyclic, up-and-down fishery as crabbing. Yet no one could have predicted such a thunderous collapse. The worst part was that nobody seemed to know exactly what had happened, or why. One theory, of course, was overfishing. Certainly, it had an impact.

Tom Casey, the stocky, bearded fisherman known both in Alaska and in Washington, D.C., as one of the key figures in getting the two-hundred-mile-limit legislation lobbied through Congress, told me, "We took the real estate [king crab] and sucked and squeezed everything we could out of it. And we felt genuinely human in doing it. Why? Because the Department of Fish and Game, and the National Marine Fisheries Service set the guidelines. They said it was okay. Hell, we fishermen weren't about to argue. We just devoted the best Yankee ingenuity to the chores. And we did it in spades."

American businessmen had used tax-free money and unparalleled technology to produce crab boats with incredible efficiency, with which nature could not possibly keep pace. These hunter-killer ships led the invasion. In the head-on collision between the *Paralithodes camtschatica* and the clever *Homo sapiens*, the crustaceans had lost, outright. Those in authority had deemed it legal, and fishermen had wasted no time in taking them up on the plan.

While the theory that we had completely overwhelmed king crab populations was being debated, we had to agree that, as with the buffalo, there was little doubt that we could have slaughtered them unto the last fleeing crab.

Fishermen pointed to the allocations set by the Department of Fish and Game. The Fish and Game turned to their biologists for the answers. The biologists needed time. The feds didn't know which way to turn.

Others pointed suspecting fingers at the record high numbers of predator fish like the halibut, skate, and cod—fish that foreign fleets once scooped up by the billions of pounds. In the few years since the foreign fleets had been kicked off the Kodiak and Bering Sea king crab grounds, there had been a virtual explosion in bottom-fish populations. There, possibly, lay the reason for the unbelievable shortage of small recruit-sized crab.

Then there was the factor of changing water temperatures. Over the past several years, the temperature of water in the Bering Sea had climbed a whopping five degrees Fahrenheit.

There was no end to the blame. Try disease, predation, and overfishing. And while one is at it, toss in the warming effect of the El Niño currents and the probability of the simultaneous arrival of a particularly low downside of the normal up-and-down fluctuations in the crab population's cycle. Likely, all played a role.

The oceanic cycles of growth and decay had always undergone change, but seldom had all the favorable environmental circumstances come together as they had in the Bering Sea during the late 1970s. Factors such as light predation by other fishes, cooler water temperatures, a near-total absence of disease, and a relatively nominal amount of fishing pressure had all aligned favorably.

In time, biologists were able to determine that a disease had spread through much of the king crab population, leaving the females sterile.

As to why the populations of king crab exploded, even more than a decade later, National Marine Fisheries biologists cannot explain the unparalleled numbers of young king crab that entered the population in the Bering Sea in the early 1970s (and later ballooned into unheard-of numbers of adult male king crab). But they do know that the absence of such crab in the years since then has played a key role in the relatively low number of harvestable king crab. Many claim that today's population is closer to the normal level.

Today, researchers generally agree that overfishing was not to blame for the sudden decline in the numbers of crab—that the collapse would have occurred regardless.

For many deckhands, the boom king crab days in Alaska had produced a fictionlike opportunity. The stories of opportunity and profit, as broke young men journeyed out into America's last frontier and struck it rich, read like chapters torn from an Horatio Alger novel.

Many of the youths in our ranks purchased salmon boats and limited-entry fishing permits, and gave up crabbing altogether. They now spent their summers gillnetting along Alaska's immense coastline, from Bristol Bay to Kodiak, and from Prince William Sound to southeast Alaska.

Others invested in real estate, building houses and renting them out, or put their money in apartment houses or bought and sold land. A number of these ex-crabbers retired before they turned thirty.

But there were those who had invested heavily in the king crab fishery who had never been to sea. They were the unseen owners, the businessmen in Anchorage and Seattle and Portland. Following the 1978, 1979, and 1980 boom years of the fishery, they had unwittingly looked to the fishery to keep expanding. They had come to expect the phenomenal annual profits, much the way they had come to depend upon the security of their other land-based investments, which had always produced a steady profit.

Fishing was never a sure thing, however. There always had been drastic fluctuations, and when the fishery collapsed, it collapsed so suddenly and completely that many investors were carried straight into financial disaster.

Following the collapse of the king crab boom, scores of crab-boat skippers chose to convert their ships into draggers. As always, though, politics was involved in lining up a Bering Sea joint-venture contract with either the Koreans, the Japanese, or the Russians. Crab boats with stern-mounted wheelhouses weren't much use when it came to dragging a mid-winter net behind them. And these crab boats went on to fish for brown king crab and opelio crab out on the edge in the lonely and savage outer reaches of Alaskan waters.

They fished the scattered populations of brown (deep-water) king crab in the savage weather and howling wind-tunnel passages around the tide-ripped waters of Adak in the western Aleutians, or fished for opelio crab (a small subspecies of tannerlike crab weighing about a pound apiece) in January on the edge of the polar ice cap (for thirty-five cents a pound). Many of them geared up for supplemental fisheries they would never before have considered, grab-bag fisheries such as long-lining for halibut and black cod, or tendering herring. And during the summers, they fought hard to land salmon-tendering contracts with the canneries, something many boats performed as favors during the boom king crab years.

Yet they found themselves faced with not only the sudden collapse of one of the most quick and profitable fisheries in U.S. history but also with new expenses. There was no way to simply park a 120-foot crab boat in the garage. Insurance premiums of $50,000 a year and more had to be paid. There were also the costs of maintenance and upkeep, not to mention that entire families depended upon the ship to continue to produce.

Diesel fuel, for instance, which only a few years before had sold for 18 cents a gallon, now sold for $1.25, which meant that a medium-sized crab boat with a former fuel bill of only $10,000 or $11,000 a year now had to shell out upward of $60,000. When the fishing was good, as it had been, the fuel overhead had always been low. Neither the 1979 nor 1980 king crab seasons had lasted a month, which kept fuel bills delightfully low. Suddenly, boats were having to fish all year long to remain financially solvent, however. A large crab boat grinding away month after month on the different new supplemental fisheries might spend in excess of $100,000 on fuel alone!

Now the board members and presidents of the banks who had loaned the money and financed the new vessels found themselves in a tight spot. In many cases, after the tight years of 1981 and 1982, they had long since thrown out their standard schedules for foreclosure. Their stockholders, and an entire fleet of crab-boat skippers and investors, were forced to sit and wait in the hope that the crab populations that had once so suddenly and profitably flourished might miraculously do so again. For once seized by court order, what could a banker do with a 120-foot king crab boat rusting dockside in Seattle? Banks carried the boat loans until all hope had been exhausted and further faith became the act of a religious fanatic.

By the winter of 1984 (after a second closed year of king crab fishing around Kodiak and in the Bering Sea), scores of the great steel vessels could be found sitting empty and idle, rusting at anchor in harbors stretching from Seattle to Homer, and from Kodiak to Dutch Harbor. The zenith of the fishery had passed, at least for the present. And as boat owners scrambled to convert their crab boats to mid-water trawlers, names of well-known pioneer fishermen who found themselves overextended and forced into Chapter XI retirement grew with each passing month.

The headlines in the *National Fisherman* magazine read FISHERMEN FACE DISASTER AS ALASKAN CRAB GROUNDS CLOSE.

Like so many of my friends, I looked upon Alaska as an alternative to the rush and hustle and crowded indifference of the South 48. Here, I believed, a man or woman with ambition, a strong back, and a little luck could still find his or her way.

There was a slip side, though. Word was leaking out, and Alaska's population was swelling. Thousands were flooding into the forty-ninth state. They came by car, boat, plane, and ferry. In Anchorage, people were arriving in mid-winter without jobs, or even the remote promise of one. Unable to afford to move their families south again, they slept for months in the backs of vans and campers and travel trailers, and when their propane gas supplies ran out and the lives of their families became threatened, they went to the city of Anchorage for help.

Local residents dubbed them the "Lost Horde," members of a modern-day gold rush (in the form of oil and construction, timber and crab) that had gone bust. With the pipeline finished and the oil flowing into Valdez, the unemployment rate in Alaska had soared to more than 30 percent in places.

The poster on the wall in the ferry terminal in Seattle read IF YOU DON'T ALREADY HAVE A JOB IN ALASKA . . . DON'T GO!"

EPILOGUE

In the summer of 1984, near the crest of the Chilkoot Pass, a four-day, forty-mile hike out of Skagway, I sat alone on the edge of a granite cataract and dangled my feet over a canyon ledge with a river roaring several hundred feet below.

As I sat there, I could picture Kodiak fisherman Mike Doyle struggling free of the crab pot sixty feet under, and the incomparable beauty of those mainland days spent aboard the *Royal Quarry* with Susey Wagner, Steve Calhoun, and Jonesy. I could feel pride at years of work well done, then recall with perfect clarity the gray elephant-like skin of the processing ship slipping narrowly past. I could picture Rusty Slayton diving for his boy, recall the world-class storm in the Bering Sea, hear Harold Pedersen's Mayday call for help, see the giant three-story waves collapsing across our deck, and recall their thundering report.

I could see the *Epic's* courageous Bob Waage adrift in the inky darkness and hear his steadfast refusal to abandon the side of his unconscious friend, and I could recall the grateful utterances Gerald Bourgeois expressed when he was rescued after being shipwrecked for eleven days. I could see my Norwegian crewmates dancing with delight on the deck of the *Williwaw Wind;* our crab pots rising full of spiny red king crab; and the enormous paychecks.

I could see the faces of the men I'd worked with at sea and partied with ashore. I could feel the "go to hell" spirit of the bar crowds in Kodiak and Dutch Harbor, hear the final vote by fishermen to settle the seven-day-old Dutch Harbor strike, and recall how the entire fleet rushed out of the harbor. And I could feel once again what it was like to plug the holds with 213,000 pounds of king crab in less than a day of fishing.

I could wander the ice fields in search of our crab pots, battle ice storms, picture the *Gemini* rolling over in an arctic storm, and see Wayne Schueffley and his two crewmates clinging to life in a drifting life raft for five continuous days and nights. And I could share their joy at being rescued.

In my mind's eye, there was Gary Wiggins and his crew leaping from the stern of the burning, sinking stern of the *City of Seattle;* and Big Steve Griffing putting it all on the line as he maneuvered in the deadly surf off Sitkinak Island. Then I thought of the panicking crew of the *St. Patrick*—young, foolish, heroic, and doomed.

There was Bob Kidd, the second survivor off the *St. Patrick,* clinging to the cliffs of Afognak Island as the giant surf stripped his survival suit from his body. I would recall forever the hope and courage of several thousand townspeople who gathered along the waterfront in Anacortes in support of the relatives and to pray for the fourteen fine young men missing off the capsized *Americus I* and *Altair II.* That and ten thousand other images of Alaska, the sea, the land, and its people swarmed in my brain.

The last I heard of the crew of the *Royal Quarry,* Susey Wagner had given up fishing and was living in Nome, Alaska, where she was finishing her apprenticeship as a certified electrician in the Prudhoe Bay oil fields. Former crewmate Steve Calhoun saved his money and went on to become co-owner—and skipper—of the *Royal Quarry.* Mike Jones expanded his holding to include ownership of a number of crab boats, and went on to become a highly successful businessman. The *Quarry,* with Steve Calhoun at the helm, continued to do well fishing and tendering around Kodiak Island. She is still one of the first boats out of the harbor and one of the last to return.

Lars Hildamar continues to run highly successful fishing vessels in Alaskan waters. Several years after I departed so suddenly from the *Williwaw Wind,* I bumped into him on the streets of Kodiak. He paused long enough to inquire as to my well-being and to shake my hand. He seemed friendly and wished me well. Robert "Bobby" Ragdé went on to fish for several more seasons aboard the *Williwaw Wind,* then seemed to drop from sight. I had not seen him in years when, only recently, while hoofing it down a street near the waterfront in Ballard (the Norwegian section of Seattle), I came upon a sign mounted above a door marking the entrance to a rather clean and dignified pub. The name of the place stopped me in my tracks. The sign read simply BOBBY'S PLACE. It was one and the same man.

As for the crew of the *Rondys,* Terry Sampson eventually gave up fishing and now runs his own landscaping business in Oak Harbor, Washington. Crewmate Dave Capri married Jill and built a fabulous home overlooking Yaquina Bay in Newport, Oregon. While Vern Hall turned his attentions to mining gold on the banks of the Yukon River (among many other business ventures), Dave became the permanent skipper of the *Rondys.* He fishes for brown king crab out at Adak in the western Aleutians.

Tom MacDonald, the skipper of the sunken *Elusive,* never did go back to crab fishing. The last time I talked with him, he was building handcrafted fishing boats down in Oregon.

As for myself, I had grown tired of the sad litany of grave-site ceremonies. I had grown tired of the senseless, endless loss of life. And I decided to take a sabbatical.

With my feet hanging over a sheer canyon ledge, and the ever-steepening trail up the Chilkoot Pass rising above me, I continued my break, filling my eyes with the wild, boulder-studded mountain country of southeast Alaska. Here the land was so steep, its character so raw and uninhibited, it touched my heart with fear when I contemplated moving over it.

Directly in front of me, a mountain rose up and disappeared into a blue-black layer of storm clouds. On the face of the sheer slope, a river poured out from under the cleft of an immense diamond blue glacier. In its steep, unrestrained rush into the canyon below, the river tumbled and fell through a jungle of brilliant green rain forest, cutting a pristine path of brilliant white as it came, all the while echoing its white-water descent. Eventually, in that cathedral of nature, what I saw lifted me entirely free of myself, until only the vision remained.

More than one hundred miles later, I arrived in Canada's Yukon Territory at the outpost of Whitehorse. Then, in preparation for the two-month journey ahead, I bid successfully on a seventeen-foot fiberglass canoe, stowed several hundred pounds of food and gear aboard her, and shoved off alone on a nineteen-hundred-mile canoe trek down the boiling powder blue headwaters of the primitive Yukon River.

In the decade since the peak of the boom king crab days out of Kodiak and Dutch Harbor, I have returned to Kodiak and the Bering Sea dozens of times to fish for halibut, black cod, opelio crab, tanner crab, and king crab in the Bering Sea now that the stock seems to be returning.

In recent years, I have found that the mammoth new high-tech hunter-killer crab boats with their modern hull designs, navigation systems, hydraulic deck cranes, and line-coiling machines have replaced much of what crewman used to do by hand on deck. However, while technology has removed much of the glamour, it seems to have removed little of the danger.

Each May, a four-day celebration called the Kodiak King Crab Festival takes place in Kodiak, Alaska. The festivities include carnival rides, a seal-skinning contest, a footrace to the top of Pillar Mountain, swimming relays in survival suits among ship crews in St. Paul Harbor, and a Blessing of the Fleet waterfront parade of local fishing boats.

The festival begins with a solemn moment, as well, for each year, regardless of the weather, crowds of friends, fishermen, and relatives gather in front of the Fishermen's Memorial Monument beside the harbormaster's office. It is a time of prayer and religious ritual, a time when the names of the local fishermen lost at sea over the past year are read aloud. As each name is recalled, a priest strikes a bell, sounding the toll.

In 1988, the bell tolled forty-two times.